CW00816271

Sonic Agency

Part of the Goldsmiths Press Sonics series

Goldsmiths Press's Sonics series considers sound as media and as material – as physical phenomenon, social vector, or source of musical affect. The series maps the diversity of thinking across the sonic landscape, from sound studies to musical performance, from sound art to the sociology of music, from historical soundscapes to digital musicology. Its publications encompass books and extensions to traditional formats that might include audio, digital, online and interactive formats. We seek to publish leading figures as well as emerging voices, by commission or by proposal.

Sonic Agency

Sound and Emergent Forms of Resistance

Brandon LaBelle

Published in 2018 by Goldsmiths Press
Goldsmiths, University of London, New Cross
London SE14 6NW

Printed and bound by Clays Ltd, St Ives plc
Distribution by The MIT Press
Cambridge, Massachusetts, and London, England

A CIP record for this book is available from the British Library

ISBN 978-1-906897-51-2 (hbk)
ISBN 978-1-906897-53-6 (ebk)

www.gold.ac.uk/goldsmiths-press

What you want to hear, you hear not. For, what finds its way out from the underground and the out there is spoken in rhythms and tones, in a language that solicits a different hearing.
(Trinh T. Minh-ha, *elsewhere, within here*)

Does not the perspective of a better future depend on something like an international community of the shaken which, ignoring state boundaries, political systems, and power blocs, standing outside the high game of traditional politics, aspiring to no titles and appointments, will seek to make a real political force out of a phenomenon so ridiculed by the technicians of power – the phenomenon of human conscience?
(Václav Havel, "Politics and Conscience")

Contents

Acknowledgments

The conceptual framework of *Sonic Agency* developed during a visiting fellowship I was granted from the Department of Computing at Goldsmiths College in 2016. In co-operation with Professor Atau Tanaka and research fellows Adam Parkinson and Alessandro Altavilla, I delivered a set of public lectures in which I outlined the topic of "the sonic agent." Through such a framework I was able to bring together a range of thoughts and materials I had been circling around for some time and to formulate the overall conceptual territory of this work. I am extremely grateful for the opportunity and the collegiate support offered by the Department. Chapter 2 was initially developed as a shorter article for the anthology *ReThinking Density: Art, Culture and Urban Practice*, published by the Academy of Fine Arts, Vienna. I'm grateful to the editors for their input on this earlier version. Additional research was carried out in preparation for a visiting lectureship I was offered at Tel Aviv University in 2015. In collaboration with Associate Professor Adi Louria-Hayon and her students I was able to develop fundamental concepts related to sound and subjectivity, as well as to relate these to social and political struggles (influenced by the immediate realities of Palestine/Israel). I am especially grateful to Adi, Professor Edwin Seroussi and his colleagues from the Hebrew University, as well as Raja Shehadeh and the Mizrahi poets. Finally, a number of trips to Chile since 2010 have led to important collaborations, particularly with curator Soledad García-Saavedra. Through our work together I was afforded many opportunities to deepen understanding of the particular cultural and political histories of the country, which has significantly informed my thinking and practice. I am greatly indebted to Soledad for her kindness and generosity.

More broadly, I've been motivated by the incredible breadth of personal encounters and collaborations that have become possible in recent years, and which I see as symptomatic of a greater cultural movement toward enhancing dialogue and work with others. Being involved in a range of both institutional and independent projects, along with the free flow of exchanges with artists and thinkers, community organizers and activists, and their initiatives, has greatly enriched my own practice. And I can say, has sharpened as well as multiplied the overall moral and creative base from which I feel myself as a member of the communities in movement. To such communities, and especially to those that never have a name, I also extend my heartfelt gratitude.

1

Unlikely Publics: On the Edge of Appearance

We are realizing more and more that a poetic emotion lies at the origin of revolutionary thought.

(Jean Genet, Letter to American Intellectuals)

The development of sound studies as a discursive field prompts questions as to what defines "sound" and, by extension, methods of its study. I'm interested to engage these concerns by positioning sound and its discourses in dialogue with contemporary struggles. In this context, I'm concerned to not only contribute to an in-depth culture of sonic thought, but to also shape such thinking by locating it against social and political realities: the figures and voices that are lifted up, negotiated, interfered with, and assembled through and by sonic means and imagination. Is there a potential embedded in sonic thought that may lend itself to contemporary struggles? What particular ethical and agentive positions or tactics may be adopted from the experiences we have of listening and being heard? Might the knowledges nurtured by a culture of sounding practices support us in approaching the conditions of personal and political crisis?

These questions begin to suggest a framework by which to extend sound studies toward the urgencies of contemporary life. Such an inquiry, I would suggest, is sustained by considering and reflecting upon what it is that sound does, how it behaves and performs, what it evokes, and the ways in which subjectivity and social formations are supported and agitated by the listening sense. In this regard, I follow from what Salomé Voegelin terms a "sonic sensibility," which she theorizes in order to craft from the heard and the unheard a range of critical ideas and perspectives. In particular, she draws out a consideration of "sound's invisible formlessness" and its capacity to upset and reorient the politics of visibility.[1] Sound and listening are subsequently put forward as a dynamic framework from which to interrogate "the surface of a visual world."[2]

Sonic Agency places particular emphasis on the social experiences and productions of sound and audition, and how a sonic sensibility may inform emancipatory practices. From the continuous flow and punctuations of the audible a range of capacities and potentialities may be found. In particular, the shifting flows of vibrancy and reverberance that often shape our interactions with the world and with others, and the ways in which speech and action are orchestrated as volumes and rhythmed as durations, along with intensities of silence and noise, these form a critical base by which to approach questions of political struggle. Subsequently, the discourse that I work to develop is one tethered to the experiences and productions that capture sound's agentive potentiality. In short, I attempt to construct a larger narrative about political life by staying close to sound and listening, underscored as an expansive relational means affording dialogical exchange, the plays of recognition and the affective processes intrinsic to finding place, as well as escape routes and new social formations beyond the strictly verbal and visible. From the tonalities and ambient sonorities, along with the soundings and voicings surrounding, one gains a range of skills and resources by which to navigate the pressures and possibilities found in daily life. In this regard, sound is mobilized as a structural base as well as speculative guide for engaging arguments about social and political struggle. This allows for reflecting upon particular historical conflicts, peace and social movements as well as the nonmovements and emancipatory practices of daily life, to give detail to an acoustics of social becoming.

As forceful movements – of rhythmic and resonant intensities, of vibrational and volumetric interruptions – sound works to unsettle and exceed arenas of visibility by relating us to the unseen, the non-represented or the not-yet-apparent; alongside spaces of appearance, and the legible visibilities often defining open discourse, the flow and force of particular tonalities and musics, silences and noises may transgress certain partitions or borders, expanding the agentive possibilities of the uncounted and the underheard. Sound may carry those that struggle by way of reverberant intensities, the vibrations as well as the echoes that pass over or around structures of dominance to embolden the voices of the few, enabling strained articulations or actions to gather momentum and to take up residence within a multiplicity of territories and languages.

For example, echoes and echoing greatly lend to practices adept at appropriating and mimicking, to give support to appearances that never quite "fit in" – that return to the dominant order and the master tongue its own performative grammars and narratives yet reshaped by an altogether different rhythm, an errant migrating repetition that may sound out alternative futures. As hidden, infra-sonic energies, vibrations may also shudder the articulated and delineated forms of sociality, cohering instead around the deep matters and shared atmospheres often supporting more intimate relations. From such a view, vibratory models of alliance and sharing often interrupt the representational codings active on that "visual surface" of particular worlds, supporting instead constructs of togetherness that may carry great social and political potential.

Working through an acoustical framework, I'm led to a deeper view onto the ways in which the life of the senses is equally a political question. In this regard, I find recourse from Audre Lorde's and bell hooks' writings, which argue for an overcoming of the "false dichotomies" of the body and mind, the spiritual and the political. Instead, they steer us toward formations of community founded upon personal life experience and the emotional knowledges that are often informing one's speech and action.[3]

This more holistic approach to issues of agency is equally at play in Frances Dyson's study of contemporary political struggles, which draws out a range of critical views from the "tones" and "noises" of our times. For Dyson, sound forms a critical vocabulary by which to confront the complexity of today's crises; from ecological to economic, social to political challenges, Dyson emphasizes the possibilities afforded by way of a sonic criticality to "move toward a shared sensibility" from which to build "sense, the common, and common sense simultaneously."[4] From such a perspective, sound and listening form a supportive base from which to nurture a broader intelligence in approaching the pervasive realities of crisis.

I share in Dyson's critical and sympathetic tone, and equally engage sound as not only a medium for supporting particular struggles, but importantly as a platform by which subjects figure themselves in and around dominant orders. I focus on sound then less as a question of specific objects or case studies, and more as a set of support structures by which one garners capacities for acting in and amongst the world. I highlight this process through the notion of "sonic agency." Sonic agency is expounded upon as a means for enabling new conceptualizations of the public sphere and expressions of emancipatory practices – to consider how particular subjects and bodies, individuals and collectivities creatively negotiate systems of domination, gaining momentum and guidance through listening and being heard, sounding and unsounding particular acoustics of assembly and resistance.

Specifically, I'm interested to consider how sonic agency may assist in the bridging of the spiritual and the political that Lorde argues for, and by extension, the intensely affective and worldly labors defining contemporary life. By drawing from experiences and conceptualizations of sound and listening as being conducive to empathy and compassion, as well as the means to break the borders of particular regimes of violence with its interruptive potential, might sonic agency enable an intensification of emancipatory practices? A set of capacities derived from sound's inherently relational force and therefore enabling of new formations of social solidarity, especially as weapons against a neoliberal logic of privatization?

In this way, I position sonic agency according to what Jacques Rancière describes as a "wrench of equality" – a social force that works to interrupt the dominant order, thereby "politicizing" power.[5] For Rancière, it is by way of "the political" that power is grounded, inserting a wrench of equality into its exercise through interventions of the uncounted and the underheard.

This is what captures my imagination: *the hearing that is the basis for an insurrectionary activity, a coming community.* My understanding of insurrection follows from Étienne Balibar's examination of modern democratic systems and the way in which he emphasizes "insurrection" as being a key foundation for such systems.[6] Balibar is critical of the view that would reduce this essential insurrectionary sensibility – what he additionally refers to as an "anarchic citizenship" – by relegating the power of self-government, and the movement of people, to that of state functions and mechanisms of dominant ordering (what Rancière also suggests through his notion of "policing"[7]). Instead, insurrection forms the heart of democratic work and the meaningfulness of being equal and free by acting in "excess" to the operations of government. As Balibar goes on to state, "this excess that cannot be controlled a priori would also be a necessary precondition for the institution of democracy, because it would permit *real* conflicts to enter into the cycle of the legitimation and delegitimation of power."[8]

In this sense, sound and listening are situated as the basis for capacities by which to nurture an insurrectionary sensibility – a potential found in the quiver of the eardrum, the strains of a voice, the vibrations and echoes that spirit new formations of social solidarity – and that may support an engagement with the complexities of contemporary life. I come to imagine this insurrectionary foundation Balibar identifies as a type of disquiet, a steady drumming that resides amidst the conditions and experiences of life with others, and that lends, through its potent animations and punctuations, to expressions of critical and creative togetherness: the making of new freedoms and responsibilities. In this regard, I'm interested in how an insurrectionary sensibility works to guide particular configurations of daily practices and exchanges, of resilience and creative resistance, and that informs through a type of general intensity the perceptions and hopes of people.

Engaging questions of emancipatory practices, I further consider how we understand the public sphere, and how *being public* is seen to carry the weight and meaningfulness of political engagement. It is precisely such emphasis and equivalence – of publicness and political action – that I'm concerned to question, so as to extend the construct of what Hannah Arendt terms "the space of appearance" as being central to political reality and its expressions. For the conditions of open public engagement are put under extreme pressure, and even criminalized, by the intensification of neoliberalism and illiberalism today. To bring forward a consideration of this reality, I've sought to turn my attention to expressions and formations of underground cultures, lyrical dissidence, radical care and civic generosity, and the ways in which people drive forward a critical disruption onto dominant orders through strategies and practices born from the depths of life. In doing so, cultures and movements of emancipatory practices place us at the *limits* of the public sphere and into the struggles of the political. It is my view that such limits show us, often through tactics of invisibility and withdrawal, of migration and weakness, through actions of collective vibration or silence, pathways toward hope and new solidarities – the joining together that occurs from "below" (as Václav Havel would say) and from which one's alienation can be made productive and meaningful. Subsequently, Balibar's notion of an "anarchic citizenry" must be extended, to integrate the presence of the asylum seeker and the refugee, the expatriated and the disenfranchised, the erased and the disappeared – "non-citizens" whose illegal status or territorial uncertainty force a disruption onto what counts as "rightful." Rather, non-citizens' movements reshape the procedures of governance by demanding new rights, often by appealing to greater understandings of the human condition. This "anarchic (non-)citizenry" produces alternative practices and principles through a fundamental condition of trespass, opening pathways by which to encourage an ethics for the transnational present.

In questioning notions of appearance and making visible, I also take inspiration and guidance from the theories of Édouard Glissant, whose ideas work to engage the pluralism inherent to life with others. In particular, I'm interested in his notion of the "opacity" of worldly contact – or the *thickness* of relations. Examining the question of language within the context of the French Antilles, and the particular effects born from colonialism onto local Caribbean culture, Glissant argues for the "multiplicity" (or "beautiful chaos") such realities produce; in short, he seeks to challenge the imperialistic dominance of the French language, for example, by way of inter-lingual forms and subjects, which wield an "implicit renunciation of an arrogant, monolingual separateness" and in support of "the temptation to participate in worldwide entanglement."[9] Accordingly, conditions of visibility and appearance – what Glissant calls "clarity and transparency" – are embedded (and "dirtied") within a greater world of "opacities"; these are the dense strata of memories and histories, conflicts and imaginaries, an *inter-lingualism* found within the density of post-colonial relations and which inflect and shape productions of meaning.

The opacity of worldly entanglement Glissant describes equally performs to infuse experiences of public life, and how we appear to one another, or not, with deep complexity. I may stand up, for example, yet my standing is secured by all those that I draw upon for support, and equally, my standing to some degree occludes and shadows those behind me; I am always seeking openings and possibilities, moving toward vistas of understanding and agency, while realizing how such possibilities or illuminations are made available through a thickness of relations and the inherent chaos of social multiplicity and conflict.

Following Glissant, opacities and thicknesses are highlighted through a frame of sound and listening in order to suggest new ways for considering public life, especially in terms of how we may give power to each other through creative resistances and acts of radical sharing. It is my argument that from an auditory position – a sonic sensibility – it becomes possible to nurture modes of engaged attention, for listening is often relating us to the depths of others, and which may extend across bodies and things, persons and places: sound is a medium enabling animate contact that, in oscillating and vibrating over and through all types of bodies and things, produces complex ecologies of matter and energy, subjects and objects. From such conditions, assemblages and conversations may be fostered based on caring and empathizing, for sound and listening are highly adept as carriers of compassion and the forcefulness of one's singularity – the intensification of affective sharing (*this sound that goes right into my body*). Along with intensities of compassion and encounter, the auditory modalities and conditions of depth and care are not without their volumes and their cultural, racial or social conflicts, for sound is never far from noise, fragmentation, capture, and the inherent potentiality of raising one's voice, shouting forth, interrupting and interfering, of being overwhelmed or even silenced. In this regard, sound is political by extending or restricting the limits of the body, in the desires and needs announced in the cry, through the care and compassion listening may yield, and in acts of rupture and fragmentation, improvisation – the rapturous and violating noises that return us to the base materialism of bare life. In shuddering the state of matter and energy, bodies and things, in working to restrain or inflect particular violence, sound is a powerful force from which we learn of the entanglement of worldly contact, one that extends from the depths of bodies and into the energetics of social formations and their politics. From these energetics, which often echo and beat beyond specific localities, the few may extend themselves into a sense for being many, to conduct any number of volumes, silences and rhythms so as to punctuate their struggles.

The complex and entangled ontology inherent to an auditory position, of sonic thought and materiality, voice and care, is, from my view, enabling for a deep and generative ethics. For instance, in listening one is situated within an extremely relational instant, one conditioned by the silence of thought (attention for the other, even of oneself – the oscillations that *sound out* an inner acoustic), and in sounding forth one may vary the conditions of that attention, to nurture and care, as well as to argue and disrupt. Sound and sounding practices may therefore function as the basis for creating and occupying a highly malleable and charged relational arena, modulating the social coordinates and territorial boundaries by which contact and conversation may unfold. Through such auditory conditions and experiences one may learn from the affective and conflictual dynamics of relations how to recognize *more* than what appears in the open.

In tandem with the work of Glissant, I'm engaged in mapping this deep ethics, and am additionally struck by what Jane Bennett terms the "energetics of ethics." In her reflections on the topic of enchantment, Bennett charts out an understanding of human agency as being bound to "nonhuman manifestations" that extend from material properties to more "energetic" and "cosmological" forces. This ultimately leads to a notion of "enchanted materialism" from which Bennett creates extremely suggestive links. She writes: "An enchanted materialism embraces the possibility that differential degrees of agency reside in the intentional self, the inherited temperament of a self, a play-drive, molecules at far-from-equilibrium states, nonhuman animals, social movements, political states, architectural forms, families and other corporate bodies, sound fields."[10] Agency, as the capacity to affect the world around us, is thus interwoven into a complex assemblage of materials and forces which, Bennett suggests, requires that one "listen" – to perceive the nuanced and ever-changing relations in which the self is always embedded. Her ethical theory is one of interconnection as well as moral responsibility, yet one that does not circle around the human self only. Rather, she reaches far and wide to encourage a sensibility tuned to the "energetics" of being in the world. A world of animations and vibrations, echoes and agitations, that embeds us within the densities and opacities to which a sonic sensibility may afford deeper engagement, for it is "through sound, through the various refrains we invent, repeat, and catch from nonhumans, [that] we receive news of the cosmic energies to which we humans are always in close, molecular proximity."[11]

What might the opacities and vibrancies, the multiplicity of global entanglement and its soundings, suggest or enable in terms of contemporary struggle? Can one craft a means of empowerment by way of sonic thought, a *listening from below*, in order to nurture the power of the unseen or the not-yet-apparent? Might sound be deployed as a weapon by way of particular tonalities and collective vibrations, a listening activism, and the force of volume, to support a culture of radical care and compassion?

Tact / And Tenderness

Hannah Arendt's notion of the political as being founded on instances where citizens gather, exposed to the "common world" found in speaking and acting in concert with each other, forms the basis for a deeply probing view onto questions of agency.[12] Arendt strives, particularly in her work *The Human Condition*, to provide a greater historical arc (based upon concepts and conditions of the Greek *polis*) to the topics of political life and public formations, grounding these in what she calls "the space of appearance," which acts as an essential framework within which "speech and action" are produced and shared. "The *polis*, properly speaking, is not the city-state in its physical location; it is the organization of the people as it arises out of acting and speaking together, and its true space lies between people living together for this purpose, no matter where they happen to be." In this sense, "the political realm" arises from between people speaking and acting together, and is formed around "the space where I appear to others as others appear to me."[13]

Arendt's construction of the political realm necessitates an understanding of "public reality" as one of exposure; as she states: "For us, appearance – something that is being seen and heard by others, as well as by ourselves – constitutes reality."[14] This bringing into appearance is fundamental to the political realm, for it is here, under the "harsh light" of being public, that matters of the common world are worked out.

The notion of the political realm, as one in which appearance is essential, has generated additional arguments that problematize this obligation of exposure. For instance, the writings of Michael Warner and Nancy Fraser, in particular, have sought to expand concepts of the political realm, and the public sphere, by specifically arguing for "counterpublics" and "subaltern counterpublics" – to challenge the implied or assumed equalities and freedoms inherent to Arendt's construction of the space between people. This space, rather, is never so simple, or free of inequalities, power struggles, prejudices, imposed silences, and deep absences or vacancies. In short, the public sphere in contemporary democracies of the West requires broader and more complicated views, which presuppose cultural diversities, ethnic minorities, linguistic multiplicities, and social inequalities always already at work in spaces between people. Fraser subsequently argues for a "subaltern counterpublic" to act as "spaces of withdrawal and regroupment" as well as "training grounds for agitational activities directed toward wider publics."[15] In this regard, the space of appearance is one shaped by additional sites and necessities – training grounds and spaces of exit – which enable a political process inherently conflictual.

Michael Warner additionally seeks to unsettle and problematize notions of the public sphere in order to address the inherent exclusionary nature of its formation; counterpublics, for Warner, are equally essential. Yet he moves further, toward a more general concept of "the stranger," for as he suggests, "reaching strangers is public discourse's primary orientation, but to make those unknown strangers into a public it must locate them as a social entity." Subsequently, "public discourse circulates, but it does so in struggle with its own conditions."[16]

Counterpublics and subaltern counterpublics move us toward a public sphere in which "speech and action" must contend not only with the matters of the common world, but equally and brutally with the mechanisms that often complicate or undermine such a world. In short, political life is directed not only at particular issues or topics, but at the increasingly tensed structures and infrastructures by which people are able to gather, to be seen and heard, and which always contain hidden agendas, secret techniques of capture, forces of prejudice, strangers and agents of policing.

Stavros Stavrides furthers such thinking by shifting understandings of public space according to practices of commoning, which work in support of "life-in-common" or what he terms "communities in movement." Such communities are constituted by being open to others and by existing upon thresholds, which is enacted through various means, including gift economies and other gestures of hospitality – practices that specifically "invite different groups or individuals to become co-producers of a common world-in-the-making."[17]

Questions of contemporary social conditions, and what may constitute public life today, as well as how the "space between" people may open up to enable formations of not only public discourse but equally gestures of joining together, become central to reconstituting the political realm. This is additionally shaped by an understanding of "speech and action" as being grounded in sounded subjectivities, where speech and hearing, voice and listening, form the essential and enduring means by which to nurture spaces between, especially when such spaces must contend with contemporary systems that impose a "privatization" to our senses and politics.

This is what captures my imagination: *the hearing that is the basis for an insurrectionary activity, a coming community.*

It is clear that the mobilization of a collective power is at stake in forming the public sphere, and in support of life with others, but what of the lost individuals and the unwanted crowds, the lonely figures and the floating subjects driven by indignation and hope, and yet who may never appear in that full exposed moment of being political? In short, what forms might being political take today when the power of people is contorted by operations and systems that are mostly never apparent or exposed, that are safeguarded behind racist and sexist mechanisms, that rely upon vague and volatile market forces, and that actively withdraw into secret arrangements and fluid networks, except in those instances when individuals make transparent, through acts of insurrection, the troubling work of governmental, militaristic or corporate agencies? It is toward such actions and conditions that I'm interested to give attention, and to extend an ear in order to attend to how speech and action may resound through particular methods and tactics, often desperate or driven by conscience, and that maneuver in and around spaces of appearance to demonstrate a force of civic engagement and political imagination. They are, extending Balibar's notion, expressions of an "anarchic (non-)citizenry," often hidden or illegal, in search of support and in need of resources. In this regard, questions of "new democratic formations" require not only a discourse of appearance and exposure, but also of borders and invisibilities, liminalities and energetics, a coalitional collaboration with the non-citizen and the dispossessed, which not only highlights work and struggle, but additionally gives way to trajectories of imagination – a futurity. As David Graeber writes: "In fact all forms of systemic violence are (among other things) assaults on the role of the imagination as a political principle, and the only way to begin to think about eliminating systematic violence is by recognizing this."[18]

It is by way of an auditory position, and the projects of anarchic (non-)citizens, that I attempt to reconsider the space of appearance. By delving into the listening sense and the potentialities of sounding practices – the refrains and the reverberations by which we latch onto the world and each other – might we ground the operations of power within an ethics of the transnational present? Might sonic agency lend to the culture of contemporary self-organizing and its concern for a common world, amplifying the shuddering vibrancy of shared joys and political imagination?

In his writings on foreign bodies, the unrepresented, and the "subaltern," Alphonso Lingis steers us toward the limits of the space of appearance, outlining an ethics defined by intensities of social encounter and empirical sensualities – what he understands as the textures of worldly contact. For example, in his book *The Community of Those Who Have Nothing in Common*, Lingis gives an extremely sensitive account of the ethical encounter when he writes:

> What recognizes the suffering of the other is a sensitivity in my hands, in my voice, and in my eyes ... moved by the movements of abandon and vulnerability of the other ... and which turns one's hands, one's dexterity into tact and tenderness.[19]

Encountering those who inhabit peripheries, Lingis is moved by the deeply embodied and tensed relations articulated by way of "the humming, buzzing, murmuring, crackling, and roaring of the world" as well as the "stammerings, quaverings, and dronings of another's voice."[20] This roaring of the world, I would suggest, gives us another indication of the space of appearance, as that arena by which "people encounter each other." Punctuated by voices that not so much articulate through a clarity of communicative speech, but rather give way to the sensuality of what it means to live and breathe in a world of entanglement and conflict, Lingis guides us into the profoundly dense conditions through which we must journey in order to arrive at "speech and action." As Lingis states: "To enter into conversation with another is to lay down one's arms and one's defenses."[21] Accordingly, one's speech and actions are deeply influenced by the impingement and the generosity of others.

What I gather from Lingis is a deepening of our sensual being and a subsequent problematizing of notions of public discourse specifically when encountering and participating in this community who have nothing in common – in other words, spaces and expressions that may *exceed* appearance as the condition of speech and action, deliberation and debate; in short, that *dirty* the illuminated center. Lingis, instead, searches for a position of vulnerability and trespass – not only expressed in those he encounters, but also, and especially, through himself. For what he produces is, in fact, a form of trespass that enables a deepening of recognition, not necessarily a representation for the unrepresented, a speaking on behalf of the subaltern, a clarifying of conflict, but rather, a narrative of encounter and of being overcome: the space between as a roaring world. In short, to speak and act precisely when language falters, overcome with noise, and with what he calls "the empirical": the textured surfaces of worldly contact that demand "tact and tenderness." What enables such recognition are practices of hesitation and generosity, mutuality and renewal, a sensitivity which edges us closer to the borders of appearance, to the uncommon and the unhomed, the silenced or the hidden. The narratives and exchanges Lingis dwells upon are given as an appeal at reforming what anchors us to particular places and particular bodies, emphasizing the necessity to sense what is truly other: the unimaginable.

In the scene of encounter, instead, we are flushed with a particular intensity whereby "the noise of our throats that fills the time it takes to convey the message communicates the noise of the things or makes the things discernible in their empirical plurality"[22] – an embodied intensity, a plurality that subsumes language within the roar of the world in which sonorities echo and resonate with the deeper tonality of place, and the flow of noises that surround gives way to surprising communication and communion. As Lingis suggests, encountering each other within the fullness of worldly contact leads less to an ideal transparent reasonableness of modern individuality; rather, (mis)translations and trespasses emerge, are produced, punctuating the "chaos and opacity" of public life.

Following Lingis, I'm interested to take seriously what he names "the community of those who have nothing in common," a phrase whose seeming paradox is suggestive precisely for considering the drive of emancipatory practices. I argue for an understanding of anarchic (non-)citizenry as a base by which to consider cultures of protest less as direct actions and more as emergent formations; the opening up of a larger field of indignation and hope in which people search for ways to participate in shaping what concerns them most.

We may find an expression of this within the "Movement of the Squares" aimed against austerity measures across Europe (and finding alliance with the Arab revolts of the same period). Across major European cities, public squares were occupied, operating as important sites of popular protest whose expressions have generated a new spirit of public life and civic engagement. For instance, in Greece, throughout the occupation of Syntagma Square in 2011–12, outraged by the lack of political say, people not only demonstrated against particular economic measures, but also instigated a culture of civic generosity and mindfulness. From self-organized initiatives that effectively spread throughout parts of the country, such as alternative food distribution networks and autonomous pedagogical platforms to creative expressions of protest on the square, demonstrations gave way to a culture of common care and work as well as joyful mutuality and festivity. One expression may be considered, found in moments when people joined together in a traditional round dance. "The Syntagma Square round dances are not a strategy used instrumentally, but stem from the political homelessness of the Outraged, who had to invent new protest forms and performances."[23] Yet the dance not only captured the general energy of public demonstration, but equally staged an affective politics based on the rhythmical sharing of an ever-expanding circle.

From an insurrectionary urgency, gestures and acts are made that force into being a heterogeneous space of social becoming, whose weakness or invisibility, whose transience or strangeness upset or elide established structures to produce what I think of as *unlikely publics.* Unlikely publics hover unsteadily and ambiguously in the open, shaping themselves within quotidian spaces and locations often between communities, languages, and even nation-states, to form volatile coalitional frameworks; that draw from resources found in collective intelligence, shared skills, popular traditions, and from the energetic knowing of the senses; that build through poor and gleaned materials a space for each other, a *collective shelter*, pulling into collaboration a diversity of people, friends, family; and that continuously shift between different agentive positions, making do through an art of survival, trespass, and tactical pleasure – unlikely publics embody the speculative and dynamic force of the anarchic (non-)citizenry today. As with subaltern counterpublics, they may withdraw, only to search for new entry points. They generate public formations that *exceed* or *fall short* of legibility, producing instead an unruly public discourse – resisting the "master narrative" of the political demand in favor of lived struggles and shared desires.[24] As such, they correspond to what Stavrides terms "communities in movement" – communities formed around commoning practices that, in producing common spaces generate not so much a "feeling of belonging to a specific closed community but a feeling of becoming part of a community that is potentially limitless."[25] I'm interested in how communities in movement reconstitute the public sphere by drawing it in and out of legibility, as well as in and around state institutions, public offices, and structures. These publics are constituted by practices that negotiate traditions of political speech and action, and as such, may only gesture toward official procedures – they are fundamentally resistant to representation ("They Can't Represent Us!"[26]). Often, as in Syntagma Square, or in the case of refugee movements across Germany, they instead resort to lyrical productions and sudden festivities, civic acts of alternative instituting, as well as the sharing of a collective speech made up of accented voices and mixed languages, finding definition through a poor acoustic. In skirting representation, they in turn resist translation. While unlikely publics are, fundamentally, weak publics, they also surprise us – that is their gift, their example. Through a range of civic gestures and creative practices, they intensify the social and political imaginary, underscoring the public sphere as that which enables a broader engagement with the material and processual dimensions of life in common.

Unlikely publics may be said to build commonality not only according to the urgencies that confront them, but through daily practices that aim to reposition such urgencies. In this regard, they necessarily wield a politics of resistance by moving ahead of or around the fixtures of power; they are not always in the open, instead unlikely publics take public life with them, distributing it into small exchanges with neighbors and co-workers, into conversations sustained through diverse networks, and through the creative labors often driving a reformulation of togetherness. In short, unlikely publics are constituted by those that cannot wait for systems and states to catch up with them – they are, instead, gaining traction through the thickness of relations, giving expression to *life in the making*. Unlikely publics therefore raise the question of governance: What new social transformations and political configurations might unlikely publics enable? How might the dramatically influential transnational flows and assemblages of power and finance be reshaped according to the new civic movements, for building earthly responsibility? Can we envision a shift in constitutional orders from the resistant cultures of today who trespass borders and languages? Following the example of the drafting of a new constitution in Iceland instigated by protest movements in 2008, what expectations might we have from an insurrectionary project, particularly in terms of instating new governmental initiatives?

I approach these questions by finding promise in the new spirit of public life that not only resists, but that also delivers onto the "shores of politics" unlikely practices of radical care. Along with the new constitution in Iceland, which remains to be ratified, in Barcelona the mayoral platform "Take Back the City" initiated by Ada Colau in 2015 (stemming from her involvement in the PAH movement),[27] is deeply suggestive for how resistance and insurrection are not only temporary flashes, but that within today's extremely resourceful and independent cultures may take root to reorganize governance.

Four Sonic Figures

The issues of emancipatory practices and new public formations are examined by way of a number of modalities or "figures." These are activated as bodies of knowledge as well as constructs from which to suggest potential tactics and ways of *being political*. Subsequently, the figures of the invisible, the overheard, the itinerant, and the weak are developed. Based on particular ontological and material conditions of sound – its unseen and temporal qualities, its interruptive and ephemeral nature – these figures are mobilized through a set of historical accounts and theoretical reflections, brought together to reflect upon how communities of resistance cohere. These are figures of dispossession as well as unexpected strength, aligning themselves with the hidden and the disenfranchised, the resilient and the creative. As such, they are understood to extend away from constructs of political dominance, redefining spaces of social intensity and critical togetherness: movements and nonmovements that work in and around systems and structures of control.

With invisibility this takes shape through questions of the disappeared and what I term an ethics *beyond the face*. The unseen quality of sound is mobilized in order to consider how invisibility may be utilized as the basis for a set of emancipatory practices. If understandings of public life and political agency are often based on making visible that which is hidden or refused entry, what formations of subjectivity and social empowerment might the disappeared, the missing, or the hidden take? Considering this question, I draw upon the notion of the acousmatic, referring to a sound whose source we do not see (and central to the field of electro-acoustic music and related cinematic practices). From such conditions, the acousmatic is emphasized as a potent operation often working to unsettle relations between sound and image, between what we see and what we hear. The invisibility of sound may recondition the space of appearance by introducing a phantasmic element (*whose voice is it that I hear?*), while providing a means or vocabulary of agency by enabling one to skirt the logic of visual capture. This is extended by considering the activist work of the militant sound collective Ultra-red, whose practice utilizes sound and listening as ways to address and support communities in conflict. Their work raises questions as to how sound and listening, as the basis for critical engagement, may suggest an alternative understanding by which to constitute the public sphere, one deeply aligned with those who do not appear or appear peripheral to dominant structures. Moving from the acousmatic as the basis for political activism, I further consider invisibility through questions of the disappeared, specifically bringing into consideration the history of the Pinochet dictatorship in Chile. This history allows for an in-depth probing of invisibility, and how under the dictatorship the reality of disappearance required a set of counter-practices often based on the power and poetics of the unseen – a "black art" aligned with magic and ghosting. Invisibility is thus put forward as a way to unsettle assumptions as to what constitutes the public sphere, and as a way to give support to emancipatory practices by emphasizing how listening may direct us toward the hidden and the uncounted, as well as the faceless.

Experiences of listening are deeply connected to the act of dialogue; conversations amongst friends and family, intimate exchanges, or those occurring between colleagues or neighbors, these are often based on face-to-face experiences in which listening is mutually shared. I stand before you, I speak, and you speak back; we experience each other in this way. Yet, within such a scene listening is also easily distracted. There are often other sounds surrounding speech and direct listening. Extending from questions of invisibility, as that which unsettles appearance and an ethics founded on the face, the topic of the overheard is posed as a second figure or modality. With the overheard, I'm interested to construct a theory of relations based on interruption and the noises that often impinge onto direct listening and the conversations we have. The overheard introduces us, instead, to the strangers surrounding us. Accordingly, I use the overheard to suggest modalities of speaking and acting based on the intensities of disruption: how one may find support through the potential of volume and practices of interference. I understand this equally as the production of an encounter, one that may extend, following notions of disorder and anarchic principles, relations with others. This is supported by drawing upon the writings of Richard Sennett and Georg Simmel, whose theories of urbanism find traction through notions of disorder, multiplicity, and strangeness. Overhearing is pursued as a logic of the contemporary global city, as well as network culture, in which attention and rootedness are always susceptible to interruption, capture, to being *unhomed*, along with the eavesdropping tendencies of technology. The apparatuses that enable global networks equally impose a new condition of linkages and assemblages, making one prone to a ceaseless pressure of relations and exposure – a strangeness always close by and which is conducive to unlikely solidarities as well as techniques of capture. What forms of social and media practices may negotiate such conditions? How might the continual exposure to alterity generated from contemporary global relations nurture and enrich earthly responsibilities? From the overheard, I attempt to construct a new sense for listening within today's unhomely environment.

Invisibility and overhearing provide the basis for relating to forms of disappearance and interruption, which encourage a deep and experimental listening, a listening into the dark and toward what lurks over one's shoulder, to the side of any conversation: the presence of the stranger and the one I may not see. Strangers and the strange, as figures that produce particular encounters - as *bodies of noise* - are given further consideration through the question of the itinerant. Sound, in moving away from a source, to circulate and propagate through environments, and through matters and bodies, is deeply linked to expressions of migration and transience. Listening, I would suggest, is often a listening *after* something or someone; it follows behind this sound that is already moving elsewhere. I may hear something, but that something is never only for me; rather, it travels, it migrates - it always leaves one in wait. The itinerant thus allows for modalities and formations of agency that explicitly unsettle borders; that trespass and that deliver particular knowledges, as well as fantasies and imaginaries, founded on leaving home or nation behind. This view is furthered by considering discourses from the Caribbean, through the history of Creole languages and what Édouard Glissant and others term "creolization."[28] Creolization is posed as a process by which colonialism, and systems of dominance, may be negotiated, founded on performative appropriations, or according to what Rastafarian and reggae practices term "reasoning" and "versioning": the bending of dominant belief systems and productions through local cultural techniques, a "black fugitivity."[29] The itinerant is ultimately a figure of the foreign, *out of place*, and as Vilém Flusser argues, one able to stitch together, from passages and journeys, an errant formation. In being out of place, the itinerant may create alien languages, an *inter-lingualism* setting into motion a poetics of echoes - punctuated in reggae's mix of *riddim* and delay, along with lyrics of redemption - and that come to pass between colonial subjects, and those scattered far and wide. These ideas and references lead to a consideration of contemporary refugee movements in Europe, and how the undocumented and the asylum seeker produce what Kim Rygiel terms "bordering solidarities."[30] From such solidarities, displacement and transience yield poignant exchanges and coalitional frameworks extending between nation-states and national identities.

The itinerant though, as a figure out of place, one constituted by being far from home, is equally a figure in need; the exile, the migrant, and the alien are always seeking shelter, searching for resources and for the means to settle; to learn new languages, to gain knowledge through experiences of the local. In short, the itinerant is also a vulnerable or weak figure. Weakness is posed as a final mode of sonic agency, for sound is never easy to hold or capture; as a material, and even as a field of study, sound is a weak object. I search for it, and yet, it is already gone; even if recorded, I must play this sound again and again in order to understand its shape and density, its frequencies as well as psychoacoustical impact. As such, it may slip through the fingers to elude description. It never stands up, rather, it evades and is therefore hard to capture fully. Weakness, though, is put forward as a position of strength; a feature whose qualities enable us to slow down and attune to vulnerable figures and the precariousness defining the human condition. Sound teaches us how to be weak, and how to use weakness as a position of strength. Accordingly, I'm interested in how sonic agency can support instances of conscientious objection, non-violent resistances, positions of pacifism and civil disobedience, all of which find strength and courage through exposed weaknesses. This is furthered by reflecting upon the counter-culture movement of the late 1960s in the United States, and in turn through a consideration of the peace struggles of East Berlin in the 1980s. Through such histories we may glimpse how refusal and resistance are often sustained and articulated by standing still and sitting in, through gestures of collective silence and attunement. Adopting a position of weakness, of vulnerability and pacifism, people find the means to resist systems of violence and domination in such a way as to highlight moral responsibility and compassion as essential guiding principles for *being political*. This is additionally shaped by what Audre Lorde terms "the erotic," which acts to articulate a "passionate politics" wherein joy and rage, sharing and giving lend to bridging the spiritual and the political.[31] From such a position one gives challenge to forces and systems that subjugate and control, to tune us toward the vulnerabilities and weaknesses all human bodies share.

Through these figures, sonic agency and unlikely publics are positioned in relation to questions of emancipatory practices. Such practices are to be understood as embodying the general vitality of life in the making – the productions and the journeys by which people figure themselves in the world. As such they unsettle and problematize directives of control and domination, searching instead for ways of being and doing that push at the seams of particular systems – that construct relations among the unnamed and the uncounted, the withdrawn and the resistant.

Accordingly, it is my concern to problematize how we understand the public sphere as often being one of visibility, serving the project of politics through actions by which we appear and relate as exposed subjects. While *being visible* is extremely important, I'm also struck by how appearance is deeply shaped by disappearance, clouded or complicated by those gone missing or silenced, or by those greatly affected by difficult journeys and intense fragility, and excluded according to a state of dispossession or disability. And further, how visibility is often skirted through practices that work to build alternative frameworks of sociality; shifting the conditions of visibility here is less about disappearing, but rather it aims at forging escape routes, secret practices, lyrical coalitions, communities in movement, and "undercommoning" cultures.[32] Sonic agency is therefore posed as a support structure for emancipatory practices, inserting into the sphere of dominant power an acoustics of social becoming and according to the rhythms and resonance that listening and being heard evoke.

Anti-political Politics

In her work on contemporary financial power, Saskia Sassen identifies what she terms "predatory formations" of financial instruments, which, as she examines, are producing new "complexities":

> Today, enormous technical and legal complexities are needed to execute what are ultimately elementary extractions. It is, to cite a few cases, the enclosure by financial firms of a country's resources and citizens' taxes, the repositioning of expanding stretches of the world as sites for extraction of resources, and the regearing of government budgets in liberal democracies away from social and workers' needs.[33]

Within these new formations, a "mix of elites and systemic capacities with finance as a key enabler" ultimately work to consolidate a high level of concentrated power and wealth.[34]

The ability to develop intense concentrations of resources and capacities is what concerns Sassen, leading to a series of examinations into the operations of financial instruments and the new logic of "expulsion" shaping them. Rather than economic development bolstering a broader middle class (within Western countries), enriching the population (as was the case following the Second World War), the new formations of financial power reduce social benefits, instead concentrating wealth increasingly into the hands of the few. While corporations excel at maximizing profits through a range of "new tools" – "advanced mathematics and communications, machines that can literally move mountains, global freedoms of movement and maneuver that allow them to ignore or intimidate national governments, and increasingly international institutions that force compliance"[35] – Western governments and central banks argue for a reduction in social programs. From such a situation, the possibilities for engaging in the movements of power through political processes are deeply reduced. As Rancière states, politics simply becomes a form of policing.[36]

Sassen's critical analysis ultimately suggests that it is within the financial arena and spheres of banking that power firmly resides, which contributes to the deterioration of political arenas and civic institutions. The "financial logics" at work today spread across the social and political spheres to situate even the essential needs of human living – "bare life" – within a matrix of profit.[37]

Subcomandante Marcos, writing from the mountains of Chiapas, describes neoliberalism as a deeply destructive system, one that extends from earlier conditions of industrial capital to take on greater force. Under the conditions of neoliberalism, he writes, "companies and states collapse in minutes, not overthrown by the storms of proletarian revolutions, but by the battering of financial hurricanes." He continues:

> Certainly neoliberalism has created a formidable enemy force in large-scale financial capital, an enemy force that can dictate wars, crashes, dictatorships, so-called "democracies", lives and above all deaths in every corner of the world. Nevertheless, this process of total globalization does not mean the inclusion of

different societies, incorporating their particularities. On the contrary, it implies the implosion of one single way of thinking: that of financial capital. ... It implies the destruction of humanity as a sociocultural collective and reconstructs it as a market place.[38]

In this context, where does the public sphere, and its related space of appearance, take shape? How do people engage with the powerful dominance of financial networks and the work of expulsion and extraction? And further, as Wendy Brown asks, "what can democratic rule mean if the economy is unharnessed by the political yet dominates it?"[39]

It is clear that a sense for political engagement has been deeply frustrated, manifesting in an intensification of social movements and occupations, grass-roots initiatives and practices of commoning and undercommoning, along with the intensification of populism and new formations of right-wing movements (given new force with Brexit and the Trump presidency). Increasingly, attempts are made to reclaim a sense for direct democracy, and to construct alternative frameworks and shared platforms against illiberalism by which collective self-determination may take shape to become operative for the benefit of a greater good.

The deeply committed and agonistic movements at work today, though, still raise the question: how might one enter into the arena of financial power to find a foothold, and to enact a form of political will? How might one confront the deeply penetrating forces that wield great influence over one's life? It seems the new concentrations of power require, if not demand, a continual probing, with ideas, imaginings, and actions – new vocabularies of gestures and social formations, perceptions and co-productions – in order to contend with what Edward Snowden has also shown: the new operative mechanisms of global instruments whose scale is truly beyond individual reach and yet which reach deep into individual well-being.

Given such a shift, it seems imperative to continue to think beyond traditional constructs of the public sphere and political work in order to articulate other tools and resources, and to extend how and in what ways appearance may take shape. It has been my task here to approach questions of politics and power by reflecting upon emancipatory practices, and in what ways speech and action may be supported through a sonic agency.

Further, how particular constructs of cohabitation and common care may gain traction through an affective politics, supporting assemblages of often disparate struggles to create solidarities across borders, to actively tune and detune particular acoustics – the flows and agitations of coalitional resonances and their intensities. Through the lessons and affordances found in a range of historical and contemporary examples of social struggle, listening is marked by its capacity to instill sensitivity for what goes unheard. Listening, as Deborah Kapchan argues, enables us to linger within "spaces of discomfort" for the benefit of exchange and dialogue.[40]

The contemporary formations of "financial predators" that Sassen examines mostly function by *disappearing* into operations of networks and digital instruments, as well as behind corporate protections and transnational conglomerates. Synthetic finances, technological formations, tax benefits, and networked systems are deeply at play in today's global environment – shaping and driving the "market place" Subcomandante Marcos identifies. In this regard, the notion of "reality" Arendt puts forth, as predicated on the work of appearance, is one that in fact never truly or unconditionally appears; instead, its designs and operations, its coordinates and territories are full of distributed networks, virtual communities, compounded financial mechanisms, and algorithmic structures whose circuits and fibers conduct and police a great deal of reality by remaining beyond one's view and strictly outside mechanisms of accountability or open dialogue. Accordingly, people must work across this new frontier, enacting an array of functions that interconnect daily life with global finance, personal desires with communication networks, educational studies with credit transactions, work with precarity. In this context, one must begin to rethink a notion of agency so as to craft additional tactics, discursive constructs, shared resources, working structures, and collective intelligences to contend with the penetrating operations and systems determining relations of power. Hence the social movements and creative projects that work to not only resist certain policies, but additionally attempt to reinsert forms of sensuality to worldly relations; that seek to construct structures of togetherness and plurality, retaking the processes and procedures of assembly, and to slow down the productions of oneself in order to figure new ways of populating a space of appearance within daily life.

Vassilis S. Tsianos and Margarita Tsomou underscore current cultures and movements of protest as "performances" aimed at reigniting the political imagination.

> Here, the ideas and concepts of anti-austerity materialized themselves into relational processes of commoning, into flows of affects, into bodies enduring together, into vis-à-vis democracies creating shifts in the "ways of doing" or the "ways of being". One could say that these assemblies of the many with their democratic practices enacted by bodies in all their vulnerability were involved in confronting power by transforming representational partitions of the visible and the sayable into "politics".[41]

It is my view that an auditory position of sounding practices and engaged listening may contribute to the new "ways of being and doing" by offering a critical route through contemporary realities. From such a position might we imagine posing an interruption onto the arena of governance promulgated by "financial predators" by pitting invisibilities against invisibilities? To craft a relational and shared body that understands appearance as one shaped and policed by virtuality, noise, and overhearing, and that trespasses particular borders through migrations that may shadow the global reach of corporate conglomerates? And further, by amplifying the conscientious objections of the precarious and the weak within the highly privatized public sphere of today?

The agentive potentialities of invisibility and overhearing, of itinerancy and that of the weak are thus given narrative so as to lead us in and around appearance, allowing us to enter certain darknesses and undergrounds, networks and communities in movement, as well as to share in the vibrancy and interruptions – the opacities and chaos: *the empirical* – of which we are always a part. In this regard, the sonic figures that I pose here are imagined as collaborators, which may suggest the means to *listen in, toward, against,* and *with,* so as to concert with the humming vitalities of others. In this way, listening is always a working through, in the moment, opening up to what may be apparent, but equally what may exist beyond the strictly seen, for "we have the capacity to hear something about the world differently through the sounding of another's perspective; we are able to be surprised by others and by our own selves."[42] In this sense, sound operates as a generative medium for keeping open the project of a new social body.

While I hesitate to imagine this conceptual construct as a pathway for fully resolving the political challenges and personal crises that stalk us today, I hope my project may incite a deepening of imagination as to how life with others may continue to find its enhancement and grounding. In this regard, I tend towards what Václav Havel calls "anti-political politics" – that fundamental and enduring base of moral responsibility and civic generosity by which people support each other and enact forms of social change for "seeking and achieving meaningful lives."[43] It is on the level of anti-political politics that I've developed these thoughts, understanding writing, as Susan Sontag states, as a project of conscience (and I would add: *poetic emotion*). It is my hope that by thinking through sound and the relational affordances of listening one may contribute to current movements of unlikely publics, suggesting a spectrum of resources for people that struggle to find voice, that long for other systems, and that currently search within the society of crisis for new methods and collaborations, and especially that may work on behalf of a *politics from below*.

Notes

1 Salomé Voegelin, *Sonic Possible Worlds: Hearing the Continuum of Sound* (New York: Bloomsbury Academic, 2014), 3.
2 Ibid.
3 See Audre Lorde, "Uses of the Erotic: The Erotic as Power," in *Sister Outsider: Essays and Speeches* (Berkeley: Crossing Press, 2007); and bell hooks, *Outlaw Culture: Resisting Representations* (New York: Routledge, 1994).
4 Frances Dyson, *The Tone of Our Times: Sound, Sense, Economy, and Ecology* (Cambridge, MA: MIT Press, 2014), 149.
5 Jacques Rancière, "Democracies against Democracy," in *Democracy in What State?* (New York: Columbia University Press, 2012), 79.
6 See Étienne Balibar, *Citizenship* (Cambridge: Polity Press, 2015).
7 Jacques Rancière, *Disagreement: Politics and Philosophy* (Minneapolis: Minnesota University Press, 1999).
8 Ibid., 96.
9 Édouard Glissant, *Poetics of Relation* (Ann Arbor: University of Michigan Press, 2010), 118.
10 Jane Bennett, *The Enchantment of Modern Life: Attachments, Crossings, and Ethics* (Princeton: Princeton University Press, 2001), 155.
11 Ibid., 168.
12 See Hannah Arendt, *The Human Condition* (Chicago: University of Chicago Press, 1998).
13 Ibid., 198.
14 Ibid., 50.

15 Nancy Fraser, "Rethinking the Public Sphere: A Contribution to the Critique of Actually Existing Democracy." *Social Text* 25/26 (1990): 68.

16 Michael Warner, *Publics and Counterpublics* (New York: Zone Books, 2002), 106.

17 Stavros Stavrides, *Common Space: The City as Commons* (London: Zed Books, 2016), 50.

18 David Graeber, *Fragments of an Anarchist Anthropology* (Chicago: Prickly Paradigm Press, 2004), 11.

19 Alphonso Lingis, *The Community of Those Who Have Nothing in Common* (Bloomington: Indiana University Press, 1994), 31.

20 Ibid., 84.

21 Ibid., 87.

22 Ibid., 90.

23 Vassilis S. Tsianos and Margarita Tsomou, "Assembling Bodies in New Ecologies of Existence: The Real Democracy Experience as Politics Beyond Representation," in Geheimagentur, Martin Jörg Schäfer, and Vassilis S. Tsianos (eds.), *The Art of Being Many: Towards a New Theory and Practice of Gathering* (Bielefeld: transcript Verlag, 2016), 87.

24 Andrew Conio, "Revolutionary: What Do You Want?," in *The Living School* (London: South London Gallery, forthcoming).

25 Stavrides, *Common Space*, 59.

26 A slogan that circulated throughout the demonstrations in Russia, 2012. See Marina Sitrin and Dario Azzellini, *They Can't Represent Us! Reinventing Democracy from Greece to Occupy* (London: Verso, 2014).

27 Ada Colau was elected mayor of Barcelona in 2015 following her involvement in Platform for People Affected by Mortgages (Plataforma de Afectados por la Hipoteca (PAH)), which worked against evictions caused by the financial crisis of 2008. As mayor, Colau has drawn up a set of strategies for enriching citizen participation, calling for greater transparency and access. For more see: https://barcelonaencomu.cat.

28 See Glissant, *Poetics of Relation*, and Jean Bernabé, Patrick Chamoiseau, and Raphaël Confiant, "In Praise of Creoleness." *Callaloo* 13(4) (Autumn 1990): 886–909.

29 See Fred Moten, *In the Break: The Aesthetics of the Black Radical Tradition* (Minneapolis: University of Minnesota Press, 2003).

30 See Kim Rygiel, in Ilker Ataç, Anna Köster-Eiserfunke, and Helge Schwiertz, "Governing through Citizenship and Citizenship from Below: An Interview with Kim Rygiel." *movements. Journal für kritische Migrations- und Grenzregimeforschung* 1(2) (2015). http://movements-journal.org/issues/02.kaempfe/02.rygiel,ataç,köster-eiserfunke,schwiertz--governing-citizenship-from-below.html (accessed March 2017).

31 See Lorde, "Uses of the Erotic."

32 See Stefano Harvey and Fred Moten, *The Undercommons: Fugitive Planning and Black Study* (New York: Minor Compositions, 2013). I draw from Harvey and Moten's notion of the "undercommons" as a theory of resistance that specifically works from within institutional systems, such as that of higher education, to undercut the smooth operations of governance through the drive of "fugitivity."

33 See Saskia Sassen, *Expulsions: Brutality and Complexity in the Global Economy* (Cambridge, MA: The Belknap Press of Harvard University Press, 2014), 15.

34 Ibid.

35 Ibid., 213.

36 See Rancière, *Disagreement.*
37 Christian Marazzi, *The Violence of Financial Capital* (Los Angeles: Semiotext(e), 2010), 40.
38 Subcomandante Insurgente Marcos, *Our Word is Our Weapon: Selected Writings* (London: Serpent's Tail, 2001), 7.
39 Wendy Brown, "We Are All Democrats Now ...," in *Democracy in What State?* (New York: Columbia University Press, 2012), 54.
40 In a workshop held under the framework of the Dirty Ear Forum, Deborah Kapchan presented her idea of "spaces of discomfort" and how listening in such spaces is vital to facilitating understanding in today's highly divided global culture. For more on the Dirty Ear Forum see: www.dirtyearforum.net.
41 Tsianos and Tsomou, *The Art of Being Many*, 80.
42 Susan Bickford, *The Dissonance of Democracy: Listening, Conflict, and Citizenship* (Ithaca, NY: Cornell University Press, 1996), 162.
43 Václav Havel, *Open Letters: Selected Writings 1965–1990* (New York: Vintage Books, 1992), 269.

2

The Invisible

Visibility is more than meets the eye. Rather, it operates as an extensive psychological and affective base by which we feel ourselves as part of the world. In this regard, visibility guides us toward having presence in and amongst others. *I feel myself being seen.* Such a feeling – that palpable sense of being sensed, witnessed as a subject acting within the social field – lends significantly to the greater experiences of personhood and the capacity to act. *I feel myself being seen, and from there I feel myself as an actant, a person with intensity; my look equally has consequence.* In being seen, I may contribute to a specific context or community, lending through a force of visual presence a particular intensity, the intensity of my singularity.

The capacity to act gained from being visible though is never so unconditional or complete; rather, in being seen one experiences the conflictual and tensed conditions of worldly contact. Accordingly, one learns to extend or contort such experiences through a greater vocabulary of intensive gestures: I extend my voice, which elaborates the appearance of oneself for instance; I write, I move – I initiate an array of expressions which perform to shape, modulate, and negotiate the conditions of seeing and being seen. In short, I am caught within the scene of being seen. Such a condition supports one's need for being experienced, for recognition, and the potentiality derived from this, while binding one to a greater mechanism of desire, obligation, and legality – to be the object of another's gaze, to be a figure within spaces of economy and labor, of national borders, to be positioned within an arena of production or control in which my visibility is key.

Visibility is positioned as being central to the sphere of the political, to what Hannah Arendt terms "the space of appearance," and further to what

Judith Butler underscores as the conditions of an ethical imperative – *to face each other*. Yet, as Arendt and Butler both remind, visibility should be understood more as a process of continual tension in which the plays of power, desire, collective will, and self-determination are constantly at work. The complexity and dynamic of visibility leads less to a fully articulated moment of pure visibility – of being able to say: I finally appear! – and more to an existential stress defined by experiences of fragmentation and anxiety, and where I must work out my visibility, to find a route in and around its particular codings. Visibility is thus a negotiation, a conflicting procedure, and a condition of unsteady powers.

The tensions of visibility are fundamentally based on the realization that appearance is defined or inflected by forces not always visible or apparent. While visibility is emphasized as central to acts of political determination, emancipatory resistance, and collective assembly, it is equally a limit that is constantly transgressed, foreclosed, reworked and trembled by conflict, trauma, desire and uncertainty. In this regard, visibility shifts according to different social and psychic labors, giving way to bodies that appear in the open and that also withdraw, becoming invisible through expressions of camouflaged subjectivity and even of self-erasure. We become skilled at exposure as well as hiding out, at times closing the door on our appearance; maybe gathering resources so as to re-enter the social field and the relational intensities of being with others. At times, we must take a break. We struggle with visibility, its intensities and its lack, its fragmentation: *can it ever be enough, this scene of appearance?* We search for a way to be a subject within particular contexts, against particular histories, intensifying relations through visible acts and quieting them at times, through removals, silences, and even negation. Our visibility is continually modulated by ourselves and by others, by particular contexts, offices, and languages, resulting in deep lessons on the powers of appearance and its absence.

The gestures and conditions inherent to the performativity of appearance parallel the ways in which institutional visibility is also modulated, performed within a struggle over transparency and access; governments and state agencies use visibility to demonstrate, through acts of display, a level of accountability and truth-telling. Yet signs of visual display may also deflect or mask other realities, particularly those of abuse. In such conditions, invisibility may carry secrets of illegal activities; it may also assist those at odds with certain state structures and forces of power. Invisibilities, as gestures of escape or survival, can become crucial, lending support to what Václav Havel terms "the hidden sphere."[1] Invisibilities may form into practices by which to counter systems of subjugation that work at controlling who is seen or heard.

Jochen Gerz and Esther Shalev-Gerz have approached questions of history and memory through a number of important monuments, often in relation to Nazi Germany and victims of the holocaust. Their *Monument against Fascism* (1986–93) in Hamburg-Harburg, for example, brought forward a complex articulation that explicitly sought to put into question the monument as an object of historical truth and commemoration. Rather than produce a sculptural form that would withstand time and daily events, the Gerzes developed an interactive, disappearing monument: a 12-meter-tall lead column that would be slowly lowered into the ground as its surfaces became marked by the names of visitors. The column became an object by which to bring the issue of fascism, and the particular histories of Nazi Germany, into the open, as a matter that would not be neatly referenced by the object but would instead demand of visitors a level of ongoing engagement.

James Young, in his essay on the "counter-monument," highlights this crucial aspect in the *Monument against Fascism*. For Young, the work successfully underscores the gesture of memorializing as one of open debate, and the need to keep unresolved the fixity of related forms. Instead, "the never to be resolved debate over which kind of memory to preserve, how to do it, in whose name, and to what end" – these questions are essential for the project of the memorial today. For Young, monuments and memorials (in the form of public sculptures) too easily close off the process of debate, acting to house a particular historical record, and the reality of victims; the Gerzes' sculpture, in contrast, "returns the burden of memory to visitors."[2]

The disappearing monument shows us the degree to which critical understanding is supported by giving recognition to what may be hidden, by relating to what exists upon the borders of illumination, overshadowed by desires for resolution or completion. In this regard, invisibility performs not only as the negative of the visible, as an antithetical modality, but equally as an urgent set of practices that may assist in the work of truth and memory. Additionally, it may function as a means for supporting communities that operate in the hidden spheres of a cultural underground, enabling one to stake out a claim onto the conditions that make being amongst others possible.

Acousmatic Voices / Listening in the Dark / Encounters beyond the Face

Might we appreciate sound as a material event that generates conditions or experiences of non-visuality? A physical movement of pressures and molecular agitations that is fundamentally invisible, or beyond the threshold of sight – that hovers within this air, or across this skin – and that accordingly is materially between energy and event, transmission and reception? While sound may occur as part of the actions of things, in the stirring of elements or in the thrust of bodies, it nonetheless *falls away* from these originating events to motion forward, or back, up and around, into numerous unseen trajectories. Is it truly possible to map a sound's restless propagations and subsequent reflections and absorptions – its effects? Sound, in this regard, puts bodies and things into motion by *extending* their reach; a literal *moving away* that, in doing so, shifts our perceptual frame from its material anchoring, its source, toward an evanescent becoming.

This invisible quality or condition may be considered a potent attribute of sound in general, one that can enable, in terms of conceptualizing and elaborating action and agency, forms of *undercover* activity or intervention within existing situations. Hence, a sonic agency founded on the invisible is extremely suggestive for tactics of secrecy: to hover in the background, to move through particular spaces with covert intent, to give challenge to the powers of identification, ocular arrest, and visual capture. If the gaze performs to often define limits, to pinpoint those who may cross lines or borders, to delimit the permissible within the social field according to what or who appears, and how, the invisible quality of sounding events or subjects may afford opportunities for not only entering spaces – *to appear* – but for finding solidarities within the dark, or upon peripheries of appearance. *What or who is there, I wonder?* Importantly, invisibility may extend precisely what or who counts, within the space of appearance, by widening the sphere of the uncountable and the inexistent as bodies that matter.

How might bodies take on the conditions of invisibility? Is it possible to reorient oneself according to the unseen or the erased, the camouflaged or the disappearing? What types of practices might invisibility enable, especially within situations of surveillance and subjugation? I want to consider these questions by engaging the notion of the acousmatic, the acousmatic being a sound whose source we do not see, and which is taken up within electro-acoustic music and cinema in order to free sound from its context, its acoustic origin; instead, we enter an arena of "sonic objects" whose density, texture, and frequency appeal to the deep listener, and the formation of a space of sonic intensity, of darkness and limited sight.

The acousmatic is fundamentally based upon conditions of the *unseen*, of not looking, or looking elsewhere, *into* sound, and locates us within spaces of shadows, dimness, a dim light, and at times, even total darkness – a listening in the dark. Within the context of electro-acoustic music (as in works by Bernard Parmegiani, François Bayle, Francisco Lopez, to name a few), darkness is often employed within concert settings in order to dislocate sound from a question of origin and to reorient the listener. Darkness is mobilized to remove us from a particular context and to reorder the sensible according to absence or erasure – a negation by which something else may emerge. There is no particular body or space to which the acousmatic sonic object is contextually bound; rather, it circulates to incite a sonic imaginary – a form of listening which accentuates sound's capacity to extend away from bodies and things, and to request from us another view onto the world, one imbued with ambiguity.

The acousmatic, from my perspective, situates us within a complex space by which recognition is shaped less through visual identification and face-to-face relation, but rather through a concentrated appeal to the listening sense. Who am I then within this space of listening, and what is my relation to others? In what way do sonic objects redefine these relationships, and how might they inform our understandings of appearance and subjectivity? Might we consider the acousmatic as the basis for a type of ethics, and even politics, one that may engage a condition I would characterize as being *beyond the face*?

Drawing upon the seminal work and theories of Pierre Schaeffer, and the field of electro-acoustic music, the militant sound collective Ultra-red provides an engaging application of invisibility inherent to the acousmatic. Starting from HIV/AIDS activist work in East Los Angeles in the early

1990s, Ultra-red came to practices of audio recording and composition as vehicles that enable forms of protection and secrecy, as well as from which to nurture community engagement and building. As the group recalls:

In order to counter police testimony in the event of arrest and prosecution, the needle exchange workers made extensive audio recordings of their work and encounters with law enforcement officials. The decision to use audio rather than video protected the anonymity of those who participated in the program. These audio recordings were incorporated into Ultra-red's first compositions and installations.[3]

Within the context of community work, audio is used as a means for navigating the threat of arrest. Yet, Ultra-red's use of audio goes further than being a strictly functional tool; their practice works through the conditions of sound, adopting methods and tactics that rely upon the experiences we have of listening and of being heard. Listening is mobilized as a productive and organizational act. In this context, it becomes not only a question of making visible communities often marginalized by social norms or abusive powers, but also of putting into question the power structures that force some to appear over others, and that require, through a type of ocular pressure, that one announce oneself into a space of appearance in order to gain access to rights and care, and to political life in general.

In other words, we move beyond the question of whether one is or is not permitted to "appear" within the public sphere to an investigation of the contestations and contradictions that produce the social geographies constituted by the operations of both visibility and invisibility.[4]

These contestations and contradictions operate as critical platforms by which to gauge the struggles aimed at shifting the limits between public and private life. To appear or to be visible is never a clear and unquestioned given, rather it is produced according to the inherent power dynamics central to the public sphere.

This is not to denigrate speech but to understand it as conditioned by those who hear and hear differently, by dissonance rather than triumphant consensus. In these struggles, "invisibility" marks the condition of possibility for the construction of analyses intended to counter the regimes of visibility that dominate the public sphere.[5]

Within Ultra-red's practice and methodologies, invisibility comes to act as a crucial device to not only bring into question the lines that keep some in place, but equally to spur the force of what I might term "horizontal" listening – a listening out for what is and is not there.

How do voices find the courage and the wherewithal to speak, especially when appearance is dangerous? And how might we hear these voices in such a way as to move beyond presumed assumptions, or the limits of dialogical arenas? Might listening enable a process of community work, especially in terms of bringing together highly contested issues and diverse subjects?

Methods of social engagement for Ultra-red importantly include processes and protocols for dissonance as well as consonance, for listening and for listening *again*, for a horizontal listening which "may be used to de-familiarize everyday sounds and voices, thereby delaying the point at which a sound's causal and referential properties are identified."[6] Within these gaps and delays one may begin to hear beyond the normative patterns that often delimit audibility and the dialogical discriminations always at play in speaking.

Ultra-red take their cues from a deep interest in the acousmatic, and the theories of Pierre Schaeffer, who importantly "describes how listening to the properties or material qualities of sound, such as volume, consistency, duration, placement in the binaural field, and texture, requires drawing the veil between the signifier of the sound object and its cause, the signified."[7] The acousmatic functions as a generative tool, a condition or operation by which to undo much of the embedded or reactive impulses that mostly support normalizing structures and that return us to what we know. Defamiliarizing our perceptions, veiling the relation between signifier and signified, asking us to listen *again*, acousmatic listening becomes a base from which to build anew relations to the social and political realities that surround particular communities. As the group suggests, "Sound recording, editing, and processing technologies also make it possible to foreground discrete and hidden elements of a soundscape, revealing qualities and resonances that may otherwise be difficult to hear."[8]

Michel Chion gives a detailed account of the acousmatic in his books, *The Voice in Cinema* and *Audio-Vision*. Stemming from an analysis of cinema, Chion characterizes the acousmatic as a "fluctuating zone" that moves in and around what we see.[9] In doing so, it may support the presentation of the visual, the scene, and actions of characters, and at the same time, it may give challenge to these elements by suspending their logic, by explicitly *haunting* the experiences we have of looking. In being an invisible presence, acousmatic sounds are highly mobile, able to relocate from on-screen to off-screen space, to change in intensity, and to inflect what we see with a psychological charge. It is precisely through a type of radical freedom – an ambiguity of meaning and intention, and of origin – that acousmatic sounds may inflect cinematic experience with degrees of uncertainty and emotional intensity.

Chion's theories of the acousmatic find recourse through a reference to psychoanalytic theories of listening. The intensities derived from the sonic elements found in film, for Chion, refer us to the more primary experiences of sound and hearing experienced in childhood. By way of example, Chion refers us to the infantile experiences of "the maternal voice," which, following what Guy Rosolato and others term "the sonorous envelope," immerses us within a "bath" of sound.[10] It is within the enveloping "womb" of the maternal voice that primary experiences of listening are to be found.

From such a view, Chion elaborates an understanding of the relation between sounds and moving images. For instance, the ways in which the mother continuously shifts between appearing and disappearing before the child, is present as a proximate body full of tactile and vocal assurance while also withdrawing, behind doors or into corridors, behind "the screen" of its perceptual world – these experiences drastically shape how one later comes to relate to and experience relations between sound and image, imbuing them with unease.

Chion suggests that these auditory experiences of intensity and rupture are replicated and employed to great effect within cinema, and are dramatically expressed through the acousmatic, or more precisely, by the "talking and acting shadow" he calls "the acousmêtre." The acousmêtre is a vague figure whose presence is expressed by a voice whose body is "not yet seen" but which promises to appear.[11] The acousmêtre therefore stages the ambiguity of sound, especially through the voice off-screen, or the voice-over, which is a voice whose image or body we do not see, or have yet to see, and that we hear as a type of phantom unsynchronized or dislocated from its image. It is a voice that we *apply* to the visual scene, and yet it holds an ambivalent relation: is this acousmêtre a form of inner voice that we, as viewers, mysteriously access? Where does it come from, this voice, and is it a voice to be trusted? In short, what kind of voice is this? – *who is really speaking?* This ambivalence is, for Chion, a potent reminder of sound's capacity to *haunt* the image and experiences of looking. The acousmêtre, he suggests, is primarily "malevolent" – *it carries secret intentions.* In this regard, the uncanny oscillations between sound and image produce a state of unease, fixing our attention within a perceptual (and familial) structure that requires continual psychic labor, a working through.

The acousmatic, in requiring of us a type of psychic labor, a negotiation with what has gone missing, or what we may not have access to, incites our imagination as well as fantasy.[12] Returning to the work of Ultra-red, listening becomes the means for deepening a view onto one's surroundings: by asking one to listen and listen *again* to recorded sounds of certain environments or events, prompted by the question "what did you hear?", one is essentially asked to hold sounds within a framework of deep attention, so as to tune us to revelations or insights into what is there or not there, who speaks or who doesn't, and what dominant tonalities are at work. Yet, the question "what did you hear?" may also incite an unsteady process of association and imagination, imbuing our listening with what we might like to hear, or what we thought we heard. In short, the acousmatic as a procedure of separating audible events from particular contexts is in itself always conducive to associative imagery, fantasies, and projections: to bring focus onto a given sound may lead to deeper knowledge and vocabulary pertaining to the sonic object, but it may equally support speculative and haunted hearings that are always linked to the unconscious. Ultra-red give a hint of this issue when, in a text related to a community project held in Dundee, they underscore how important it is to "foster a condition for listening that binds us together through anecdote, story, memories, joking but often through the pain of expression."[13] The nurturing of such a particular space, one inflected by the personalities and subject positions of participants, the pains and the pleasures, becomes necessary for moving forward with collective work and struggle. In this regard, Ultra-red remind of the importance of starting with what we all bring into the space of listening, which must include not only the capacity to intervene within particular community politics as informed subjects, but to do so through deploying one's faculty of imagination and even fantasy. This is certainly part of what makes the acousmatic so enticing, for as Chion suggests, in haunting what we see or what is apparent with a particular uneasiness, it requires of us a range of perceptual and creative powers. The acousmatic may ultimately tether us to the psychic tensions of desire and meaning, while providing a vehicle for relating to what lies beyond what we see, the hidden sphere of particular powers and repressions, demanding a shift of the rational and the reasonable; in short, to oscillate as sounds do across multiple modes of knowing and sensing.

I would extend Chion's acousmêtre figure toward the larger field of sound and hearing in general, to suggest that the experiences of auditory events do much to puncture our psychological constructs with continual intensity; in short, audition is lived as a process of constant agitation, a fluctuation by which we learn of the temporality and ephemerality of bodies and things. Sound is never permanent or immutable; rather, it carries the conditions of ambiguity and fluctuation, as a force of oscillation that requires of us continual psychic labor: to find or construct meaningful points of support through the pleasures of hearing while navigating the ruptures and fragmentation the audible imparts or produces.

From the topic of the acousmatic and the plays of voice in cinema, I'm particularly interested in how the voice-over, as an acousmatic sound occurring off-screen, acts to *trouble* the face and its appearance, shaping or masking it through conditions of ubiquity, omnipresence, and fluctuation: the voice-over seems to know more than the figure on the screen; it has access to spaces and knowledge that exist beyond the frame. In this regard, the acousmêtre is intimately linked to fantasy and trauma, and the rending experiences that shape subjectivity. If, as Chion suggests, the primary experiences of sound we have as a child work to intensify the temporality of the world around us, how we hear or what we allow ourselves to hear is defined by a deeper psychological matrix shaped by the piercing lessons of dependency and ambiguity, repression and desire.

The acousmatic can be understood to reinforce the material intensities of things around us, while relating this to the spectralities of what is missing or held behind the "screen" of the real; it incites a type of psychic labor aimed at recovering or retrieving a particular body (the body of the mother, perhaps, which may represent a sense of wholeness or security). In doing so, the acousmatic performs as a *hinge*, placing us between the uncertainties intrinsic to hearing and the work needed to wrestle with the ongoing drama of the seen, for the acousmatic educates us on how every image is inflected by what it shows as well as conceals. From such conditions, the acousmatic produces a form of affective knowledge, one that acts as the basis for a "listening activism" – a listening that may intervene within systems of visual capture.

We may further consider these ideas by following Chion's analysis of the film *The Invisible Man* (1933).[14] Based on H. G. Wells' novel from 1897, the film gives narrative to Griffin, a scientist researching optics whose penchant for unconventional experimentation leads eventually to the fabrication of a particular concoction, a potion that, after repeated consumption, slowly induces a condition of invisibility.

The disappearance of this man leads to meditations on the human condition and forms the basis for a critical appraisal of the existential struggles of being a body amongst others. Invisibility provides a unique opportunity for exposing the normative conditions that enable human subjectivity and social contact; in other words, visibility is exposed as the core means by which one gains agency and affirmation as well as legal status: as the film often suggests, the invisible man poses an inherent threat to law and order through his inability to be counted within the social field and, importantly, to be arrested. The invisible man is ontologically criminal.

The invisible man gives stark expression to the acousmêtre mapped by Chion, for this is a man whose voice is never stable; it is a voice losing its body, disintegrating from the screen to take up a position off-screen, forever dislocated from its image – we hear him speak, but there is no reliable face, no mouth or figure to which it is linked. In fact, his body is only made available through clothing and the mysterious bandages he wraps around his head – the bandages enable him to "appear" before others while marking him as "wounded" – a gaping wound – thereby enhancing and even exaggerating his alien presence. This disappeared subject.

These ruptures are played out as challenges to the social order and the conditions of the sensible, giving way to the "madness" that lurks within the folds and fantasies of hearing: subjectivity shaped by the traumatic memories that imbue the heard with degrees of ambiguity, pain, and risk. Invisibility is threatening because it frees this man from the ordering of the social and the legal which are based upon visibility. Instead, he becomes nothing but sound, fixing him to the untrustworthiness and unreliability of the acousmatic, this sound we hear but that never returns to us its image. This is played out on a number of levels. First, there is the removal of origin – invisibility causes this man to be nowhere and everywhere at the same time: he is essentially a stranger (he even speaks with an accent); he is dislodged and without homeland (he arrives in town seeking shelter), and is therefore always a figure of the unknown. Secondly, the sense of omnipresence, and of having supernatural power – as Griffin himself proclaims, an invisible man may rule the world (invisibility produces a type of unimaginable power: it may overtake everything, relegating the world to nothingness, to the void as empty as this man's gaping absence: what if the invisibility inflicting this man were contagious?). Finally, the operations of the power of ghosting – there is a terror central to this man who disappears, one that I may highlight as being connected to a notion of radical freedom: might the invisible man haunt us precisely because he is beyond the constraints of moral conscience, structures of law and society, beyond the face and its incorporation or requirement within the space of appearance; a space of ethics and of politics, surely, which is fundamentally a space of accountability? The invisible man is, in short, uncountable. In this regard, he draws out the anxieties inherent to audition, utilizing them against us, as well as exciting the imagination, enabling us to fantasize all sorts of possibilities.

* * *

The invisibility expressed by Wells in the character of Griffin is extended and problematized by Ralph Ellison, with his own novel of the same name published in the early 1950s. Ellison gives us another rendering of invisibility, in this case through the lens of racial prejudice in American society. Politicizing the figure of the invisible man by giving a stark depiction of the dispossession enacted through racial abuse, turning people into second-rate citizens according to the norms of white society, Ellison equally locates invisibility as a complex platform of desperate existence as well as social transformation. It is through the state of invisibility that the unnamed main character works out a mode of speech and action, one that leads to terrible confrontations as well as uncertain pathways for recuperating one's dignity.

The invisibility that Ellison captures highlights the ways in which visibility is not always already one of agency and empowerment; Ellison's main character is in fact seen by others – he *does* appear in a range of social situations – but in being seen he is equally erased, overlooked, or seen as having no consequence onto the deeper shape of public life. In short, his invisibility is enacted by the brutalities of a certain visibility: he appears yet without a face.

Ellison's novel stages the deep complexity of visibility as being central to self-determination, of recognition and its empowering and reassuring touch, as well as its ability to dispossess another, withdrawing another's feelings of recognition through the casting of a look that ultimately disregards. The invisible man is a figure withdrawn by white society and brought into his "proper" place by the power of a look that denigrates – that hates what it sees. This is captured poignantly in a scene where the main character attempts to break his unendurable situation: throwing himself into the face of a white man, he shouts and hollers, pleading to be seen, asking of the other to recognize the ethical demand his face should carry.

Returning to Griffin, we might understand how he is equally caught: while he attempts to disappear, he is always held back by the look of others, by the eyes that refuse his disappearance. While Ellison's character searches for ways to have a face, Wells' invisible man fantasies about losing his. Griffin searches in the territories of the withdrawn and the missing for a truth: the truth of the absence every image and body carries. Through such a situation, Wells shows us the challenge invisibility poses, and how looking does in fact impose upon others a type of demand: the demand to stay tied to the social order and according to particular power dynamics. Both characters, through their particular journeys, give challenge to the social order, one by demanding recognition, the other by attempting to thwart it.

Invisibility leads to insights and revelations about bodies and embodiment, about agency and social action, and the traumas and ethics of disappearance. To be without visual representation, while deeply undermining social and political engagement, is counter-balanced by what invisibility may also provide: the conditions for occupying the limits of the normative structures by which political subjectivity and social work are made meaningful. Invisibility may enable means for escape and withdrawal, for covert intervention and antagonism, as well as for survival. To be without face thus poses a range of breaks and cuts onto the social order, bringing forward any number of anxieties as well as forcing into action experiments in radical freedom.

The Black Arts / And Black Readings / From Histories of Secrets / (Im)possible

The question of disappearance is not without complexity; rather than a mere act of losing substance, fading into an ethereal domain of airy being, disappearance may in turn carry great weight, to remain lodged within the individual body in the form of deep memory and the presence of what refuses resolution. As a coordinate within the conditions of visibility, disappearance poses a problematic, one bound to experiences of loss, fragmentation, and even death.

Disappearance is equally a tragedy of great proportions, one conducted through actions that force certain bodies into oblivion. Disappearance is an unresolved death, where mourning is thwarted, thereby suspending the emotional logic of life lived and grieved by others into a perennial question: *where is his or her body?* How might I grieve this person whom I have known, and even loved, and which is now nowhere to be found? And to which no one is held to answer for? Such disappearances refuse to go away, and instead, contour the world of appearances with deep unease.

Subsequently, we enter a space of emotional intensity, one that requires a form of negotiation as well as acts of remembrance, yet often according to formulations and gestures not necessarily ordered by way of a rational logic. Instead, disappearance – this body that I know is out there and yet never fully appears – requires a deviating practice, practices of fantasy and incorporation, as well as that of magic and superstition, where rational understanding and reasoning shift to alternative modalities (modalities that are always on the border to psychosis: *I cannot stop looking for what has gone missing; it invades my body – I promise never to forget*). For the disappeared occupy a nebulous and difficult arena that pushes us to the peripheries of reasonable thought, of politics and social dialogue, and of emotional ordering. How to account for this body that has disappeared and that remains perennially obscured even by explanations that are often never fully truthful or clarifying?

Invisibilities play out through practices that gain traction from alternative logics, those in which illumination and transparency converse with the unseen, and through intuitions, poetics, and magic (which may be called "madness" from the normalizing perspective of the rational). Magic is positioned here as a platform by which one gains insight onto the unseen from the acousmatic movements of sonic events, through spectral witnessing, and the performances of what I term "the black arts." *How else may I live with this phantom body, this body that occupies my own?* I give it a shape, a secret name; I tether it to something (an emblem, a totem, a photograph) by which I may address the unsteady presence of this death unresolved. The black arts are an art of shadowy constructs which deal in the dark knowledges emerging from violence and trauma, and that dialogue with the unspeakable.

I'm interested to expand upon the acousmatic as the basis for addressing the disappeared and the greater question of visibility. From the interweave of critical knowledge and imaginative projection, conscious knowing and the unconscious associations at play around seeing and listening, the acousmatic provides a productive channel. Might we apply the acousmatic to the tragedy of the disappeared, to train one's listening for approaching and maneuvering through the emptiness of the seen? To listen again to the absence that refuses to be silenced and yet which cannot truly resound in the open? In what way can the acousmatic enable a type of "reading" through a reorientation of the senses, one able to contend with the silences of the withdrawn?

Voluspa Jarpa, an artist working in Chile, leads us into a deeper understanding of disappearance and the formulation of related practices. Contending with the legacy of the Pinochet dictatorship, and the thousands of disappeared and tortured to which the country is still deeply bound, Jarpa has devised a number of aesthetic strategies that enable a form of remembrance while problematizing the notion of illuminated truth and recuperation. As is well known, the Central Intelligence Agency of the United States played a pivotal role in the overthrow of Salvador Allende in September 1973 in Santiago de Chile, and the subsequent dictatorship of August Pinochet (1973–89).[15] Through this period, and prior, the Agency actively monitored, consulted, and documented governmental offices and their activities, resulting in an archive of extremely crucial documents; in fact, the CIA documents provide an important view onto the periods prior to and during the dictatorship, and give insight onto the otherwise secretive operations of Pinochet. Deemed "classified" until their public release between 1999 and 2000 (under President Bill Clinton), the CIA archive was to give access to important information, in particular to the status of the disappeared and those responsible. Yet the documents were mostly censored, with large sections redacted. In this way, the release of the archive both enabled a certain historical accounting while exacerbating such account through its partial telling and brutal deletions (which only compounds the disappeared with additional violence). The blacked-out, redacted, and deleted sections therefore come to form a type of record, yet *in the negative*.

Voluspa Jarpa has taken up the highly charged obfuscation of the documents, producing a number of important works, all of which give expression to the complex experience and condition of disappearance as that which is never so clearly absent or withdrawn, but whose nagging persistence unsettles the objective of history and its telling.

> The redactions suppressed the texts, inverting the letter into image, presenting enigmatic voids between phrases and images, and what's more, revealing the traces of secret information. In that sense, Jarpa's works have taken up those images that are and that represent the negation of history or, failing that, those images that are and represent histories of secrets.[16]

As the curator Soledad García-Saavedra suggests, secret information seems to require another type of interpretation, a dilation of focus that may assist in apprehending what remains locked behind blacked-out paragraphs. The archive, in this way, necessitates a form of contemporary activation, a particular practice in order to negotiate the violence of the withdrawn; the violence continually at work around the disappeared body and compounded by the appearance of a censored document – a document whose materiality stands in for those who remain buried in the Atacama desert, or that lie at the bottom of the sea having been thrown from helicopters by military agents. They are there, we know, and yet such knowing is only ever a spectral force, a vaporous formulation discussed undercover, a cryptic or watery knowledge whose meaning and interpretation finds recourse through a black art - an art that may assist in reading blacked-out paragraphs. Secrecy thus breeds secrecy, the disappeared speaking through deleted passages, and the words marked by partial truths.

One such instance of black reading is found in Jarpa's work *La Biblioteca de la NO-Historia* (2010–12). The work consists of a set of bound books each containing a selection of the declassified CIA documents and ordered chronologically. Of the 22,000 documents released, Jarpa focuses on a selection of 10,000 which are, as she suggests, "reclassified" through her bibliographic assembly. Jarpa, in a sense, re-enacts the documents' partial withdrawal by publishing their obscure pages, staging their declassified release, and in doing so, enables an unexpected form of reading, however tenuous or ambiguous. The printed books ask us to contend with their blacked-out pages as records nonetheless, whose negativity and withdrawal incite a reading not only *between the lines*, but also *into the black*.

The act of such readings is not so much based on a forensic analysis aimed at the reconstitution of a truth, but rather a poetical transformation, one that relocates the missing into a logic defined by absence and the phantasmic, and one that may shift analysis toward speculation, illumination toward compounded obscurity. Lingering upon the threshold of communication, the disappeared articulate a "language of haunting" that defines the ambiguous zone between the living and the dead.[17] As such, a black art may assist in relating to the disappeared, to the ghost always present in the archive. In short, we move into a zone of aesthetic engagement whereby "the impossibility of transmission"[18] is given a type of material shape, not so much through recounting personal biographies of those gone missing, but rather through a phantasmal re-appearance - a circulation of secrets, muted memories, and blacked-out pages from which the disappeared come to appear.

The disappeared body is transmuted into an archival body, a dark library, one that stands in public space and demands our gaze; yet the library is resistant, the blacked-out paragraphs and deleted lines deflect my anxious look. I am left on its surface, and there, if I stare long enough,

I may begin to detect the appearance of the disappeared as a threshold, a periphery: a limit. Here, an ethics *beyond the face* is required – I attend to what I cannot possibly grasp but which I know exists in its *inexistence*; even the faces that once appeared as photographs of those missing pasted across city walls in Santiago, even these are but faded traces that struggle to surface from within their own historical void. In short, they become markers of a perennial absence, an absence to which one must nonetheless stare blankly, in search for what may lie beyond its muted surface, a depth that may only be accessed by way of secret passages, black readings, a bewitched reasoning. Jarpa's work ultimately "makes visible a material that puts visibility itself into question."[19]

The difficult reality of the disappeared is not necessarily resolved through this library of blacked-out pages and its public appearance; an act of forensic analysis, through a probing of surfaces, scraps, informational detritus, does little to resolve the circulation of individual accounts, shared narratives, rumors, and whispered tellings. Rather, the disappeared continue upon their phantasmic course, threading their way into narratives passed across the country whose words and speeches, whose cultural artifacts carry, and care for, the emptiness of what resists retrieval.

Emmanuel Levinas' "ethics of the face" may act as a philosophical partner when confronting the reality of the absent and the missing, as well as the haunting potentiality of the acousmêtre. Levinas insists that it is by way of the face that one confronts the human condition in all its stark demand, and from which a transcendent power is inaugurated: the power that binds one to the other, and from which acts of violence and murder – a killing of the other – may be overcome. Accordingly, the other's face, in disrupting me, holds my subjectivity against an ethical limit: I cannot turn away from this face that impinges upon my person. Without even speaking, the face of the other calls to us, and obliges us to pause.

How might Levinas' ethics of the face perform in relation to the disappeared and the violence intrinsic to it? What forms of responsibility can be employed to counter state-sanctioned terror, which perpetuates itself precisely by removing the faces of those that oppose the state? A negation that in turn eliminates the faces that may bear witness by equally thwarting processes of public truth-telling and accountability? If there is no body, can there be a trial? If, as Levinas poses, "the face is the evidence that makes evidence possible,"[20] how do processes of justice and truth-telling result from faces that specifically refuse entry as evidence, and whose disappearance is perpetrated so as to nullify such processes? A telling that may be performed by those that have witnessed what otherwise should not be seen?

Acts of violent disappearance and withdrawal are based upon such negations, thwarting processes of witnessing and grieving, as well as the operations of an ethics that, once confronted with the face of those held captive, incites a clear response. In contrast, within this absence of appearance, in the secrecy of a nothingness, one is haunted by what cannot be retrieved and therefore spoken of fully; we are left only with remnants, partial features, *invisibilities,* the dust of pictures and objects, clues that stage their own inadequacy and to which we must construct a form or expression of responsibility and understanding. In short, we are left with the silence of the disappeared and the ambiguities of partial tellings, the sounds of those that attempt to speak on behalf of the missing face, often through secret codes and uncertain accounts, and whose words and actions compound the tenuousness of knowing; that give us an equally ghostly outline by tracing what otherwise remains invisible - a vague register standing in for the missing body. Such voices and actions are driven by an ethical limit - by the obligation for the other whose face I do not see; an ethic thwarted along the way, left hanging and expressed in gestures of hope as well as deviating practices of interpretation and projection - *through its absence I know this other exists.*

Avery F. Gordon, in her work on ghostly matters, equally considers the question of the disappeared, in this case within the context of the dictatorship in Argentina (1976–83). The haunted languages of the disappeared that Gordon considers lead her to demand a type of accountability as well as the courage to speak openly and directly about those gone missing. As she writes: "To withstand and defy its haunting power requires speaking to [the disappeared] directly, not paralyzed with fear, out of a concern for justice."[21] The ability to speak directly to or of the disappeared though is never truly achieved; rather, in the case of Argentina and of Chile, the disappeared are never fully recoverable or made openly addressable. Rather, such histories of violence extend beyond any single case to create a culture imbued with haunting absence, perennially raising the question as to how one goes about approaching that which "transgresses the distinction between the living and the dead" – the absent friends and family members that linger everywhere and nowhere, and that generate a cavity within the project of collective memory. What type of communication may we then use to "speak directly," and can such communication ever be free of the fear and power of that which refuses to go away? Acts of communication instead veer into shadowed routes, at times bypassing language altogether, and drawing from a range of spectral and acousmatic channels. In this regard, one brings forward the labors of one's imagination, which may assist in negotiating not only the brutality of the disappeared, but also one's own projection cast onto the darkness left behind. Here, the lessons of the acousmatic provide a general framework by which to contend with the absences pervading the seen; as a black art, the acousmatic traffics in the missing, enabling us to penetrate the surfaces and representations of the social order, to learn of what often goes unnoticed, overlooked, the repressed and the forgotten. In this way, the acousmatic acts as a vehicle for figuring the dark into a source of knowledge, as something we may hear.

The aesthetic operation at work in Jarpa's library is one that forces one to look *again*; as with Ultra-red's invitation to listen, and to listen *again*, as one that may pose a rupture onto the normalizing responses of the cultured ear, these blacked-out books stage their own limits as well as possibilities. It is precisely on the threshold of a perceptual attentiveness – the limit of the graspable – that one may hear and see beyond oneself, to formulate, for a moment, the possibility of *knowing otherwise*, and from which to relate explicitly to what has been withdrawn from the world. In this sense, a black art is a craft aligned with magic and a logic of ambiguities and liminalities, of transactions that communicate through oblique messages, temporary meanings, and truths defined by spectral forces – by the unspeakable around which we are drawn.[22]

We may additionally explore the topics of history and memory, and the performative operations of the missing as found in Chile, through another example, this time one staged by President Ricardo Lagos in 2003. The particular event I want to consider was organized to commemorate the thirtieth anniversary of the military coup, and included the dramatic action in which President Lagos ceremonially walks out of the front entrance of the Presidential Palace, around the side of the building, and through a re-installed side door, the door through which Allende's dead body was taken out during the coup 30 years earlier. Following the coup, this door was removed, blocked, withdrawn from public view and access; it was as if it never existed, and by extension, that Allende's presidency was only a phantasmic event, one that may take on mythical proportions while also disappearing into the void of history. Such disappearances, over time, may lead citizens to pause, and wonder: *did it ever in fact exist?*

Pinochet not only became a practitioner of brutal violence, but one whose power derived equally through gestures of trickery and what Michael Taussig terms "the magic of the state." Drawing from his long-term ethnographic work in Colombia, Taussig poses magic as being tied to paradox and the circulation of a spirit of mystical exchange, one that may lend significant input into state economies and military cultures as well as expressions of popular unrest and resistance. As he outlines, "It's as if state and people are bound to the immanence of an immense circle of magically reversible force, in effect a never-ending exchange"[23] across which forms of signifying articulations and counter-articulations pass. The "magic of the state" is therefore a highly volatile and deeply charged channel through which relations of power are contoured by a continually oscillating logic, one aligned with spectrality and fluctuating meanings. Within the magic of the state, the acousmêtre can be found.

Pinochet's particular form of state magic was based on the art of persecution, yet one masked by economic policy; one hand withdraws the power of people into rooms of torture and graves of sand and water, while the other performs tricks taught by the Chicago Boys on how to initiate economic reform for the benefit of the nation. Money and death were thus two sides of a violent ethos that would demand equally magical and painful forms of resistance. One sphere of resistances is to be found in the creative practices of Diamela Eltit, Carlos Leppe, Elías Adasme, and Raúl Zurita, among others. Whether through the self-inflicted cutting of one's flesh or the use of bandages as materials within performance, or in the act of burning one's cheek or hanging oneself upside down on the streets of Santiago, these actions attested to the systemic violence of everyday life in Chile during the dictatorship, while also indicating a set of strategies for maneuvering through and making sense of its continual pressure, its haunting presence – a presence that could, at any point, appear from out of nowhere to take one behind the veil of the withdrawn and the disappeared. These artists and writers produced works that lead into the dizzying terrain Taussig describes, in which the violence of state activity is counter-balanced by self-induced pain; according to an "immense circle of magically reversible force," persecution is negotiated through gestures of confusion and masquerade, trickery and the production of obscure meanings; a black art designed to smuggle dissent, and to generate vocabularies by which to nurture cultures of resistance – performances whose aesthetics of wounds and lacerations, of excess and obscurity, self-abuse and self-negation, of compounded fractures as well as fractured poetics sought to counter the dark armies of the dictatorship. The production of veiled meanings served to clothe one's wounds and to give shelter to the lost nation, forming a cloak of secrecy by which to aid voices as they attempted to speak – poems in honor of one's own imminent disappearance. Removals and erasures became brutal currency by which politics and economics, artistic practices and social resistances were thus performed.

Raúl Zurita's extended poetic work, *Purgatory*, maps this uneven relation, where existential conditions became a scene of dark analysis and conceptual dissection: a set of lyrical passages that spiral upon themselves, a purgatory in which bodies are prodded and poked for their weaknesses. Within such liminal states, the material of flesh and bone, blood and breath, are tensed by a condition of anguish.

i. Let's look then at the Desert of Atacama
ii. Let's look at our loneliness in the desert

So that desolate before these forms the landscape becomes
a cross extended over Chile and the loneliness of my form
then sees the redemption of the other forms: my own
Redemption in the Desert

iii. Then who would speak of the redemption of my form
iv. Who would tell of the desert's loneliness

So that my form begins to touch your form and your form
that other form like that until all of Chile is nothing but
one form with open arms: a long form crowned with thorns

v. Then the Cross will be nothing but the opening arms
of my form
vi. We will then be the Crown of Thorns in the Desert
vii. Then nailed form to form like a Cross
extended over Chile we will have seen forever
the Final Solitary Breath of the Desert of Atacama[24]

Zurita's purgatory is not solely that of a journey to the end of the night; rather, his poetical view is shaped by slow persecution and the insidiousness of a state apparatus as it inflects language with a logic of abuse, of wounds and capture. In this context, Zurita's poetics is wielded to aid in the plight of vulnerable bodies, crafted so as to give structure and cadence to the terror of systemic violence, figuring it as the basis for a resignifying attempt, a meaningful construct.

A similar aesthetic attitude can be found in the works of Carlos Leppe, for example, in which performances appeared as constructed scenes of death as well as possibilities for a type of transcendence; or in the work of Diamela Eltit, whose performance *Zona de dolor* (Zone of Pain) (1980), for example, also stages the complexity of systemic violence.[25] The performance unfolds as a series of gestures, the first of which is based on the artist cutting and burning her own arms. This act of self-abuse is subsequently captured on video; shot by the artist's collaborator, Lotty Rosenfeld, the video slowly scans over the artist's body as she sits with her arms hanging loosely over her knees; she stares into the camera, she displays her bloody arms. This section is followed by the artist entering a nearby brothel where she reads aloud parts of her novel *Lumpérica* to the women and men gathered there. The conflation of her acts and the context of the brothel locates the performance in relation to a question of vulnerable bodies. In this case, we occupy a zone of sexual exchange, one hidden behind closed doors and under cover of the night, which is equally a zone shaped by military curfew. Finally, following this second section, the artist finally exits, returning to the street outside where she begins to wash clean the stones and pavement in front of the brothel.

As with Zurita's *Purgatory, Zona de dolor* locates us within a series of confrontations and rituals related to bodies and their vulnerability. Eltit seeks out the edges to the state structure, bringing her pain into alignment with spaces of secrecy and nocturnal gathering. Hers is an impoverished, lyrical ode to the body subjugated by a greater state magic and against which one conducts one's own performance, producing a scene of pain infused with lyrics of lament and obscure meanings, a spectral construction wherein pain is turned into mystical surrender.

It is within these zones of pain and purgatories that resistances were sought, echoing with what David Graeber, in an analysis of communities in revolt, refers to as the "creative reservoir of revolutionary change." As he reflects: "It's precisely from these invisible spaces – invisible, most of all, to power – when the potential for insurrection, and the extraordinary social creativity that seems to emerge out of nowhere in revolutionary moments, actually comes."[26]

The walk of President Lagos in 2003 thus revolves upon an axis perennially aligned with wounds, disappearance, and rituals needed to address the deep cavity of the nation. While searching for ways to heal the gaps and breaks of the country, his walk can be seen to participate within the ongoing narratives and spectral cultures founded on disappearance. The President steps outside to walk around the Presidential Palace and back in through the once deleted door, to perform a symbolic re-appearance of the original President, the one we may never retrieve and yet whose lingering memory is still very much alive to act as a narrative of the broken nation, a figure equally standing in for the bodies of those never found.

Returning to the scene of the acousmatic, and Chion's cinematic analysis, I'm tempted to recast this mysterious reappearing door found on the side of the Presidential Palace as a cinematic screen; in walking through its once erased passage, what Lagos attempts to recover is precisely the missing "maternal" President Allende – to break the "seen" always haunted by the acousmatic voice so as to recuperate the disappeared, the nation that never was.

Disappearance functions as a deeply historical question, yet one that haunts the present through the complex force of absence. How to write and read such history? By which language might such negative force be inscribed, and according to what evidence? As the President walks back into the Palace, a type of re-enactment occurs, though in reverse: he does not come out of the door, as the body of Allende did, but rather, he enters back in; a gesture that both attempts to heal, to literally force back upon itself the haunting legacy of Pinochet, as well as one that re-opens the

Palace to the possibility of a future to come, a future defined by a body able to challenge the force of disappearance – by retracing certain steps, deepening their negative presence even while attempting to redraw them. In this regard, we may glimpse the degrees to which disappearance and the missing define the visible world, and in this case, the narrative of citizenry. As the artist Luis Guerra suggests, President Lagos, in this instance, performs the role of the "first citizen," a body free upon the streets of Santiago and one that makes an allusion to Allende's final statement, a voice broadcast over radio – this acousmatic voice still haunting the nation – as the bombs fell upon the Palace, that one day in the future man will again walk freely in the nation of Chile.[27]

The Hidden / Yet Heard / As New Consciousness / The Making of Faces

Expressions of visibility and invisibility, of the seen and the disappeared, are thus played out in the context of Chile in such a way as to complicate the illuminating project of truth. Rather, truth is to be found through a continual oscillation, producing a weave of vague gestures and articulations that move in and out of the dark. Darkness and the erased become keys to a labyrinth of understanding, where a poetics of violence and transcendence, the symbolic walk of a president, and the opening of a once closed door participate in a greater expression – not one of disclosure and illumination, but rather of meandering readings and listenings into the deep complexity of history and politics.

Questions of visibility, and the tensions inherent to spaces of appearance, are greatly exacerbated when placed within systems of totalitarian control. In response, invisibility may become a question of survival. Ivan Jirous, a Czech poet and manager of the band The Plastic People of the Universe active in the 1970s within the underground scene in Prague, theorized "the underground" as the basis for what he termed "second culture." For Jirous the underground provides an important spiritual home for dissident thought and communities of resistance, acting as a "declaration of war."[28] From such a declaration, cultural expressions come to exist *under* those defined by the establishment, which in the 70s in Prague was clearly shaped by the totalitarian system of the Soviet Union. Through "second culture" the "real aim is to overcome the hopeless feeling that it is of no use to try anything and show that it is possible to do a lot, but only for those who are willing to act and who ask little for themselves, but instead care a lot for others."[29]

In the 1970s and 80s in Czechoslovakia the underground was replete with acousmatic intensities, where questions of censorship and the potential of arrest dramatically shaped what could and could not be said or sung, and that removed or persecuted those who stood outside the law. This led to social gatherings in the countryside, with rock music and theatrical performances, as well as hidden sessions of secret listening in which albums smuggled in the from the West were played undercover. As Jirous highlights, music functioned as a key material for giving expression to "dissident" thought and from which people found the means for living through the totalitarian structures of the state. Within the musical cultures of the country at this time, bands such as The Plastic People of the Universe, DG 307, Žabí hlen, and others, drew from the psychedelic rock music of the West, such as the Fugs and the Velvet Underground, while integrating their own specific references and musical sensibilities. Freedoms were subsequently expressed through a musical world of dizzying psychedelia, often incorporating mystical lyrics that might be heard as vehicles for nurturing the second culture described by Jirous, a lyrical poetics by which to define alternative social structures and collective imagination.

Visibility is a complex political tussle that forces into relation the powers of public resistance and those of social control (which is always based on limiting the visibility of some over others). The psychic, ethical, and physical labors required to appear, to challenge, or support the appearance of others - to take responsibility - greatly define the politics of visibility, lending a palpable urgency to seeing and being seen. In this regard, appearance is less a point of arrival or moment of completion, of fulfillment or certitude, and more a continual struggle for conditions; it is a declaration requiring constant animation and affirmation.

Accordingly, practices of acousmatic listening and invisibilities, of underground sounds and black readings, give challenge to the often insistent ways in which political acts and public life are understood by way of appearance. Such an insistence amounts to an obligation to be seen, which places the possibilities of action and agency, forms of practice and thinking, firmly within a rubric of exposure and illumination - that whatever does not appear, in fact, does not exist. In contrast, visibility is cast more as a limit whose demarcation is never stable, and which we may never assume remains bound to the articulation of truth. To confront this limit, to know beyond what appears, requires additional thinking and acting, gestures that, in aligning themselves with the blindspots and the occluded, with the blacked-out and the underground, may nurture practices and positions of critical attention and creative attunement. Notions of agency founded on such positions may contribute not only to methods of attention for others, but equally a set of tactics for undermining the ways in which powers define us through our visibility or disappearance. The reverberations of the underground, the acousmatic force of a voice in the dark, and the blacked-out silences through which the hidden and the disappeared may uncannily speak – these are properties and capacities by which to craft forms of public life against powers that seek to violently hold us within particular limits. Accordingly, they generate a complex and multi-dimensional framework, in which rational knowing and critical inquiry gain traction through capacities and resources often found in spectrality, associative and underground knowledge, fantasy, the black arts, and magic.

Gloria Anzaldúa, a Chicana cultural and queer theorist, offers an important critical view when questioning issues of visibility, and in particular, the social dynamics inaugurated by the face. In arguing for methods of "divergent thinking" so as to counter the dominant structures of white society in America, Anzaldúa theorizes a "*new mestiza*" which "copes by developing a tolerance for contradictions, a tolerance for ambiguity." The *new mestiza* "learns to be an Indian in Mexican culture, to be Mexican from an Anglo point of view. She learns to juggle cultures. She has a plural personality, she operates in a pluralistic mode – nothing is thrust out, the good, the bad and the ugly, nothing rejected, nothing abandoned. Not only does she sustain contradictions, she turns the ambivalence into something else."[30] The skills and capacities adopted by the *new mestiza*, performed according to a plural personality, for Anzaldúa, are based on a form of psychic work, a labor that takes place "underground" she says, in the depths of the unconscious and which finds a route in and around white society through a process of "making faces." As she writes: "We rip out the stitches, expose the multi-layered 'inner faces,' attempting to confront and oust the internalized oppression embedded in them, and remake anew both inner and outer faces."[31] *Making faces* acts as a process of creatively piecing together one's identity in such a way as to combat the range of prejudices and constraints placed upon it; to perform the multiple countenances, languages, and pluralism inherent to the *new mestiza*.

Following Anzaldúa, the visibility of the face is a never-ending process of negotiation and performativity; an interplay between multiple guises, and between the inner and outer countenances particular persons acquire through living under racial prejudice; and according to the desire to celebrate "plural personality" as the basis for giving challenge to dominant codes often based on notions of the clear and the illuminated, the pure and the proper. In contrast, Anzaldúa leads us into a labyrinth of underground work in which nothing is rejected; a construct of "make believe" in which all the deeply complex pains and pleasures of oneself may serve the project of a radical ambiguity, the shape-shifting appearances adopted as means for survival as well as the critical joys of new consciousness.

Anzaldúa's plural personality, of which making faces is an act of both disguise and liberation, and Zurita's anguished poetics punctuated by the self-abusive scarring of his face through the pouring of acid upon his cheeks in anticipation of state persecution – the power the face is called upon to wield is at the same time resisted and captured for oneself according to a self-constructed logic in which disappearance is deeply operative. It is not a question of undermining the importance of appearance, and the necessity to support conditions of visibility – for it is clear that visibility is central to a confirmation of personhood, and to what may be achieved in spaces of public life. Rather, in following the ways in which certain subjects negotiate the demands of appearance, forming practices upon the fringes of visibility, I'm keen to supplement the overarching value placed upon visibility with a sense for all that exists outside the experiences and territories of the seen. Practices of invisibility start to reveal arenas in which self-determination, collective assembly, and powers of resistance and imagination are significantly played out. Within such conditions, we may detect not only cultures that challenge the status quo, and that search for means to overcome the limits of visibility, but that also steer us toward epistemologies grounded in a *negative aesthetics*, a poetics of shadows and cuts, of occlusion and erasure, and which generate other forms of signification and ethics.

Might expressions and meanings of the black arts suggest emergent forms of resistance, as well as possibilities for an ethics based not so much on this face that I see, but those I do not? An ethics that enables one to engage those gone missing or that hide in the dark, listening to music late into the night, or that occupy peripheries as means for survival? I would emphasize such figures and gatherings as *being public* yet in their negative intensity – a public of the withdrawn and the underground, a second under the first, and from which expressions of social imagination and manifestation are to be found. As unlikely publics, they wield a social and political force by shadowing the open space with obscure meanings, by making faces, producing an anguished and persistent discourse that captures the lived experiences of oppression. As such, they turn us toward the limits of the seen, reinventing the conditions that make life possible. Within these peripheries, and according to an ethics tuned to the missing, the audibility of speaking out often shifts to that of whispered and occluded reverberations and dark volumes, acousmatic sonorities which demand another form of listening – a strained, horizontal listening by which to collaborate with the unseen.

Notes

1 See Václav Havel, "The Power of the Powerless," in Václav Havel et al., *The Power of the Powerless: Citizens Against the State in Central-Eastern Europe*, ed. John Keane (Armonk, NY: M. E. Sharpe, 1990), 23–96.

2 James Young, "The Counter-Monument: Memory against Itself in Germany Today." *Critical Inquiry* 18 (Winter, 1992): 276.

3 Ultra-red, "In/visibility and the Conditions of Collective Listening," in *The Invisible Seminar* (Bergen: Institute of Art, University of Bergen, forthcoming).

4 Ibid.

5 Ibid.

6 Ibid.

7 Ibid.

8 Ibid.

9 Michel Chion, *The Voice in Cinema* (New York: Columbia University Press, 1999), 22.

10 Guy Rosolato, quoted in Kaja Silverman, *The Acoustic Mirror: The Female Voice in Psychoanalysis and Cinema* (Bloomington: Indiana University Press, 1988), 80. The notion of the sonorous envelope is also central to reflections of sound and voice in the work of Didier Anzieu.

11 Chion, *The Voice in Cinema*, 21.

12 Kaja Silverman gives a critical account of "voice in cinema," focusing on the gender politics at play in cinema as well as its theorization. For Silverman, it is imperative to reflect

upon the power dynamics found within cinematic performances, which fill the screen with a range of problematic depictions. In particular, Silverman concentrates on voice and audition, and how the female voice is positioned or constructed in relation to male voices and male listeners. Bringing her critical view onto the writings of Chion, among others, Silverman questions the degree to which the acousmatic is "fantasized" as representing a "maternal presence"; the ways in which listening is shaped by the primary scenes of childhood, of infantile experiences, are ultimately a "projection of fantasy" she suggests – one that is doubled in cinema, cast by theorists' imagination and psychic viewpoints. For who is to truly know how an infant experiences a mother's voice? In doing so, Silverman criticizes Chion for unreflectively playing out such maternal fantasies, thereby problematically shaping theories of cinema through a particular gendering. See Silverman, *The Acoustic Mirror.*

13 Ultra-red, *Dundee* [2011], 46. From the self-published notebook series, *Nine Workbooks, 2010–2014.* www.ultrared.org.

14 Michel Chion, *Audio-Vision: Sound on Screen* (New York: Columbia University Press, 1994), 126–28.

15 For more on the CIA's involvement in Chile, see The Church Committee Report and the Hinchey Report, published in *U.S. Covert Actions by the Central Intelligence Agency in Chile* (Rockville, MD: Arc Manor, 2008).

16 Soledad García-Saavedra, "Secret Archives: Suspension and Overflow in The Library of No-History," in Cristián Gómez-Moya (ed.), *Human Rights, Copy Rights: Visual Archives in the Age of Declassification* (Santiago: Museum of Contemporary Art/University of Chile, 2013), 112.

17 Avery F. Gordon, *Ghostly Matters: Haunting and the Sociological Imagination* (Minneapolis: University of Minnesota Press, 2008), 127.

18 Adriana Valdés, "Voluspa Jarpa: Biblioteca de la NO-Historia de Chile," in *Dislocación: Cultural Location and Identity in Times of Globalization* (Ostfildern, Germany: Hatje Cantz Verlag, 2011), 143.

19 Ibid.

20 Emmanuel Levinas, *Totality and Infinity: An Essay on Exteriority* (Pittsburgh: Duquesne University Press, 1998), 204.

21 Gordon, *Ghostly Matters*, 113.

22 Maurice Merleau-Ponty eloquently reminds us that the world is both revealed *and* masked by our perception of it. In motioning toward something, in touching this object, such touching gives way to a form of knowing, yet one always already partial: each touch is a type of knowing, of and in the moment, and yet remains tethered to a certain limit. One is held within a set of experiences and their memories, and yet there is always a beyond to one's knowing, which is present in each instant of experience. Knowing is therefore defined by what appears, as well as by what disappears, or is held, withdrawn, as a possibility. Knowledge is nurtured equally through a capacity to glimpse the peripheries any critical examination is surrounded by. In this way, Jarpa's work stages the archive as a limit, an incomplete collection by which to intuit what remains hidden in its pages; it is already more and less of itself, a surface of withdrawal, and yet which I hold in my hands nonetheless; an object that contains within itself the pressing forcefulness of the missing body, and which catapults my look into a maze of interpretation, into a reading still to come. Visibility is thus always already based upon its own limit. See Maurice

Merleau-Ponty, *The Visible and the Invisible* (Evanston, IL: Northwestern University Press, 1968).

23 Michael Taussig, *The Magic of the State* (New York: Routledge, 1997), 124.

24 Raúl Zurita, *Purgatory: Bilingual Edition* (Berkeley: University of California Press, 2009), 51.

25 For an important analysis of Diamela Eltit's work, and the general theme of "zones of pain," see Mara Polgovsky Ezcurra, "Zona De Dolor: Body and Mysticism in Diamela Eltit's Video-Performance Art." *Journal of Latin American Cultural Studies: Travesia* 21(4) (2012): 517–33.

26 David Graeber, *Fragments of an Anarchist Anthropology* (Chicago: Prickly Paradigm Press, 2004), 34.

27 From an informal conversation with the author, 2016. Luis Guerra has approached similar themes through his artistic and theoretical works; in particular, a performance installation staged at the former Animal Gallery, in Santiago, in 2004, titled "Fantasma," delves into the complicated reality of disappearance and memory. For more on this work, and questions of ghosting, see Soledad García-Saavedra, "Thunder Stealers: From Phantasmatic Apparitions to Ghostly Bodies in Chile," in *The Invisible Seminar* (Bergen: Institute of Art, University of Bergen, forthcoming).

28 Ivan Jirous, *A Report on the Third Czech Musical Revival* (Amsterdam: Second Culture Press, 2015 (pamphlet, originally pub. 1975)), 15.

29 Ibid., 17.

30 Gloria Anzaldúa, "La conciencia de la mestiza: Towards a New Consciousness," in Gloria Anzaldúa (ed.), *Making Face, Making Soul: Haciendo Caras. Creative and Critical Perspectives by Women of Color* (San Francisco: Aunt Lute Foundation Books, 1990), 379.

31 Gloria Anzaldúa, "Haciendo caras, una entrada," in Anzaldúa (ed.), *Making Face, Making Soul*, xvi.

3

The Overheard

The issue of sonic agency, as I'm engaging here, is envisioned as a framework for inspiring, nurturing, and empowering political subjectivity – to craft from an auditory imagination, and the experiences and promises generated from listening and being heard, emergent forms of resistance as well as compassion and care. Finding guidance according to the conditions of sounded events and their material behaviors, sonic agency may give support through capacities of circulation and itinerancy, intensities of volume and of silence, collective solidarities formed through vibration and reverberation, as well as from the spectralities and potentialities found in invisibility and the evanescent. From such capacities, might new formations of public power be considered, those that drive forward agentive possibilities? The joining together in the spaces opened up between one another through sound and listening?

There is often that sense of there being more to what I am hearing; I may concentrate, I may draw my attention toward this sound here, yet often there is something else: an excess, a background, or a push of energy that stirs below or around hearing, and yet which I know, or intuit, as *being present*. In fact, it is precisely this *more*, this background which often influences or affects the quality of what I hear by interfering with or supporting the oscillations of a certain sound: particles that touch, meet or push against another, to cancel or prolong the force and signification of what I hear. Sounds upon sounds; the overheard upon the heard.

If we consider sound as a phenomenon that hovers between an object and its propagating reach, a rather ambiguous position by which matters of the world animate themselves – flinging themselves into all directions – what it may provide in terms of guiding bodies then are means for *intensifying relations*. By passing between things and bodies, subjects and objects, sound affords an extensive possibility for contact and conversation.

In her book *Vibrant Matter*, Jane Bennett argues for a deeper understanding of what constitutes "bodies of force." Shifting from an overarching anthropocentric world-view, she instead argues (extending from Deleuze and Guattari) for life as "a restless activeness"[1] - a "vitality" that is harnessed into singular bodies, while equally expanding those bodies into a dynamic relationality. In other words, life is envisioned as a forceful capacity from which bodies are constituted, transformed, and extended. This vibratory model unsettles the human body as a defining figure, as a dominating figuring, and instead, places it within a field of life-forces from which alliances, conflicts, interminglings, and contaminations are continually at play, from body to body, species to species, thing to thing.

Accordingly, Bennett highlights agency as "distributive and confederate"[2] and that performs through "matter-energy assemblages." As she states, "In this strange, *vital* materialism, there is no point of pure stillness, no indivisible atom that is not aquiver with virtual force."[3] Instead, everything bristles with potential.

It is my view that sound, as an intensity that moves objects and bodies into the world, extending their reach and relation, is a force that works to link across singularities. In enabling an animate flow to pass between bodies and things, sound is fundamentally a vibrant matter, one that conducts any number of contacts and conversations - the rustlings and stirrings by which listening and being heard takes place. The limits of bodies and things are radically extended through sounded actions, making of them expressive flows open to intersections and overlaps, as well as fragmentations and ruptures. Sound *intensifies* relations by animating their potentiality, exposing the matters and bodies of the world to each other. Yet, these relations are also prone to being formed through interruptions and agitations - the extended and animate reach of sounded events are necessarily rapturous and disruptive; they are punctuations onto the plane of presence and within the conditions of relationships. In this regard, I'm keen to bring notions of vibrancy and their related matters into conversation with the issue of sonic agency, to consider how contemporary assemblages reformulate concepts of subjectivity by displacing its fixity and amplifying the vulnerabilities that make it open to capture and collaboration.

Suely Rolnik gives articulation to the vibrancy Bennett speaks of by concentrating it specifically into a notion of contemporary subjectivity, or what she terms a "vibratile body":

> Today's subjectivities: grabbed from the soil, they have the gift of ubiquity – they fluctuate at the mercy of the mutable connections of desire with flows from all places and times that all pass simultaneously through electronic waves. A singular and fluid filter of this immense and also fluid ocean. With no name or permanent address, without identity: metamorphosing modulations in an endless process, tirelessly managed day after day.[4]

Rolnik's descriptive image of "today's subjectivities" is suggestive for understanding vibrant matters not only as the basis for potential alliances and assemblages, but also as the condition of loss and fragmentation, as well as estrangement – she continues:

> Estrangement takes charge of the scene; it's impossible to tame it: destabilized, displaced, discomforted, disoriented, lost in time and space – it's as if we were all "homeless". Not without a concrete home (level zero of survival basic conditions where a larger and larger contingent of human beings find themselves), but without the "at home" of a feeling of oneself, a subjective, palpable consistency – familiarity of certain relationships with the world, certain ways of life, certain shared meanings, a certain belief. The whole globalized humanity lacks this kind of house, invisible but no less real.[5]

Adrift in time and space, untethered from any grounded logic, subjectivities are cast into a globalized "homelessness."

For Rolnik, the pervasive contingency of today's subjectivity is one of rupture, a type of splintering from which new "vectors of power" are created. These are formed according to an "exposure to alterity" and the persistent confrontations and exchanges that define and shape our global conditions. Exposed to a ceaseless flow of "homeless" subjects, one creates from such conditions possibilities and productions; vectors of power arising from the force of people driven into social productions, and from which new "pulsating vitalities" arise.[6] In this regard, the "homelessness" Rolnik outlines is more akin to what Homi Bhabha terms the "unhomely," which unsettles the borders of home with a troubling estrangement. "To be unhomed is not to be homeless, nor can the 'unhomely' be easily accommodated in that familiar division of social life into private and public spheres." Rather, as Bhabha continues, "The unhomely moment creeps up on you stealthily as your own shadow," uncovering what was always already there yet hidden.[7] The "unhomed" subject is, in short, defined by the structural conditions of a modernity that fundamentally unsettles origins, infusing constructs of home and identity with fragmentation and systemic rupture.

As in the work of Bennett, Rolnik identifies within global culture a redefinition of bodies and subjects driven by the intensification of relations: one is forced into ever-new configurations and constellations. With these new formations, things break apart, languages fragment, and the meanings that once cohered in the form of localities and common identities lessen, or are relocated within a framework of virtual force - or what Rolnik also terms the "nomadism of desire." This intensification or pulsating vitality may constitute new formations and assemblages, to function as "homes"; yet these are always already conditioned by flows and ruptures, by desire's ceaseless restlessness.[8] "Matter-energy assemblages" and "vectors of power" are thus contingent upon an intensification of exposure and a nomadism of desire, to produce subjectivities at once diffused and allied as vibrant matters.

The potential of a sounded, voiced, and vibratile subject extends what I can do by also breaking my body apart; in extending this body that acts, that expresses my subjectivity as an intensity of fragments and vectors, I am also made vulnerable to conditions of being *overheard.* In articulating a body in pieces, *unhomed from identity,* one is potentially captured in so many ways - in short, one is picked up, tracked and hacked, monitored and registered, followed and arrested through conditions of vibratility. This necessarily relies upon a shift in corporeality in general. Under the new conditions of global capitalism, matter-energy assemblages and life as a "restless activeness" are performed increasingly through "cognitive labor"; the new materialism of global culture redefines what constitutes matter by relating it increasingly to modes of "intelligence" and "agency" that no longer necessarily require organic bodies. In this context, bodies and minds are not necessarily housed within a stable figure, but rather are stretched into an array of vectors of power and desire.

Marshall McLuhan's notion of an extended nervous system, theorized in the early 1960s with the emergence of "electronic culture," while importantly describing forms and channels of broader global relations, is equally insightful for signaling how relations within a network are exposed to new systems of monitoring. In the thick of communications, I am also an *overheard* subject, a subject made available to any number of agencies, and in turn, I overhear. Assemblages may provide forms of distributed agency, but they are also at times the result of being corralled.

I may link up beyond myself, in the throes of matter-energy assemblages – I may become more than myself through this nomadism of desire, this cognitive capitalism – yet, in doing so, I am additionally located within any number of structures and portals of control. In this regard, vibrant matters do not necessarily escape from ordering principles, algorithmic codes, and socio-technical structures inherent to global culture. Rather, they are fueled by a greater system of virtual force in which vibrancy is harnessed into a new political economy of energies and currents, attention and mediation – a biopolitics of the cognitive subject.

* * *

Ideas of vibrant matters and pulsating vitalities are extremely suggestive for querying the agency and actions that I'm interested to engage. As oscillations moving through a range of mediums, sounds and sounded actions are vibratory in nature, and they perform as intensely relational events. Vibrancy and vitality also lead us into reflections on contemporary global culture and the subjectivities that are produced by it – this contemporary vital materialism is rather unthinkable without the paradigmatic shift under the push of digital networks. In such conditions, what happens to listening? What forms of political life and cultures of resistance are generated or prodded into being from the perspective of network environments? In thinking through these models of vibrant matters, I'm led to understand them as informative for relations between sonic agency and public power. As Jane Bennett argues, vibrant matters are fundamentally about the limits of the political, giving way to an intensity of relations that is continually at work today, and which makes a claim onto how we come to live and share this world. How then do today's *unhomed subjectivities* hear each other?

Overhearing is positioned as a narrative about the ways in which listening and being heard are transformed by the conditions of global culture. Drawing from theories of vibrancy, and further, to understandings of the stranger and urban noise, I'm interested in how overhearing may describe the relational intensities of the vibrant and the virtual, and which may assist in critically engaging both the possibilities and pressures of digital networks.

Overhearing is shaped by conditions and experiences of noise, interruption, and capture; *what I say is never only for whom I face within a zone of proximity*. Rather, my saying, in extending myself as a vibratile figure, bringing me into the enriching and volatile thrusts of multiplicity – the communicative flows and penetrating bombardment inherent to contemporary networks – necessarily places me within territories and relations I may never understand let alone glimpse. My presence is registered by numerous (robotic) agents and surveyed according to a plethora of systems of which I am never aware. In such a state, social relations become operational, *laborious*, flushed by an endless navigational desire and array of expressivities that are no longer only communicative, but are equally designed to give registration to "being present."

In this regard, overhearing is understood as a condition of listening and of being heard that integrates the complexity of the new social intensity and its relational surveying. Accordingly, proximity is always already unsettled, linked to other actants and agents, forms and forces of not only open sociality, but also technologies of surveillance, the eavesdropping of certain agents, as well as by witnesses that may inadvertently and importantly overhear, to retell or repost. Listening, in this context, is more an intersection of hearings and overhearings – even *mishearings* or *unhearings* – defined by the potency and politics of network life.

What might the overheard teach us? What kinds of lessons may we take from the force of interruption that is the overheard – is there another form of listening to be considered, one tuned to intrusion, to the intruder itself? Might a sonic agency of the overheard open up to the conditions of multiplicity as a thrust of continual differentiation? To operate as an interruption through which bodies confront bodies and by which networks are occupied?

I pose the overheard as a new general condition of hearing in the age of global life, one fully shaped by networks and digital involvement.

Following the vibrant models posed by Bennett and Rolnik, which are models of unhomed subjectivities and potent assemblages, the experiences we have of listening and of being heard are recast into a condition of ceaseless involvement and virtual interaction; the algorithmic force of the network positions one's body within a framework whose scale is perennially beyond one's reach. Instead, contact and conversation become operations around which many unseen forces are tuned, and which pressure one's attention into forms of participation that are never complete or finished. Rather, one is prodded into endlessly extending oneself – to assemble into multiple linkages, which is first and foremost a new type of labor, producing a new type of subject, a vibrant, cognitive, and overheard subject.

Extended and involved, how far can my attention truly go? The scale and intensity of the virtual force of the network riddles one's body with a vibrancy that ultimately confronts new intensities of alterity. I am forever exposed to an alterity, to a strangeness that may enrich and enliven life experiences, and at the same time requires another formation – the production of a commonality, a "shelter." How to construct an ethics of engagement within such a dizzying maze of involvement and intrusion? How to ground social intensities in local experiences and concerns? In this context, listening and being heard require another set of critical coordinates.

By bringing the background forward, delivering strangers and the strange into proximity, and locating attention within a politic economy of mediation, overhearing is posed as a model for understanding "today's subjectivities" as those defined by networks and vibrant matters – who inhabit and negotiate the conditions of multiplicity and alterity, and the nomadism at work in the psychic and affective project of digital media. An agency of the overheard, I'd suggest, may enable acts of forceful entry, to invade the scene, through gestures of interruption, and in doing so, it may also support coalitional frameworks and assemblies by which to retune the grounds on which bodies struggle; to construct a vibrant assemblage of social care and compassion.

The overheard requires us to hear differently: to find meaning in the incoherent fragments and noises that interrupt and that trouble and excite the borders between oneself and another. In being something or someone – a voice, a vitality, a murmur, a cacophony – that is unexpected, the overheard surprises by reminding us of those who are always around or nearby, behind or to the side, next door or within optical fibers: a "figure" whose agency is founded on the potential of interruption, to estrange the heard with a type of noise: with what may form into something, but not yet. Overhearing is thus always a potential: a forceful intrusion from which to escape, to resist, as well as discover new friends, tribal affiliates, collaborators, and even new love. *The stranger is always a carrier of erotic intensity*. Whereas invisibility ghosts our listening, offering routes through the inscriptions of seeing and being seen, the overheard interrupts, to turn us toward the other. *This other that is always nearby.*

Urban Scenes / A Logic of Encounters / Stranger Relations

In following Bennett's notion of "vibrant matter" and Rolnik's "vibratile body," I'm interested to consider in what ways these notions locate subjectivity within a network of intensities. The forceful relationality inherent to the distributed agency Bennett poses is one that locates us within constellations of forces and matters, energies that lend to the constitution of assemblages and coalitions, linkages that force one into interdependencies and cohabitations. I am, in a sense, reliant upon these assemblages; they are biological formations, which sustain themselves through the vitality of my body enmeshed with others. In this context, vibrant matters are explicitly embedded within biopolitical systems, as they implicate my corporeal vitality within them.

In being always already projected into a greater framework of vibrant matters, subjectivity is defined according to a state of interruption: I want something, I demand something, I speak and I listen, I motion and gesture in such a way as to tremble this assemblage of force; I am, as an intensity, an interference to others. Such a model of subjectivity, as Judith Butler additionally argues, underscores the ways in which as bodies we are always in need of others; in short, interdependencies and cohabitations are inherent to the human condition and the experience of being a body in the world.[9]

Returning to Bennett, and her meditations on vibrant matters as a "political ecology of things," we may glimpse this state of interdependency by considering our relation to food. While it is clear that our bodies depend upon food, for Bennett such a relation is deeply suggestive for human to nonhuman interaction. "In the eating encounter, all bodies are shown to be but temporary congealments of a materiality that is a process of becoming, a hustle and flow punctuated by sedimentation and substance."[10] Accordingly, eating reveals us as bodies *in process*, one that rises and falls according to the powers of foodstuff and the "mutual transformations" that occur between my body and what I consume. Eating, therefore, is "the formation of an assemblage of human and nonhuman elements, all of which bear some agentive capacity."[11]

Following this vibratile diagram, and the inherent interdependencies and cohabitations they reveal, subjectivity is in a state of constant relation that spans the social and the biological, matter and energy. The ordering of the body is thus held in balance through a process of continual contact and exchange, forcing into view an understanding of agency as one of coalition and affiliation, as well as vulnerability.

Within such a model, I am myself only in so far as I am interrupted and supported by others; this body is dependent upon the blurred boundaries that make it possible to sustain itself, while always already placing it at risk. Putting things into my mouth, I am susceptible to any number of actants. I must open myself in order to survive. In this regard, interruption and the invasiveness of the outside is a condition of life in general. At the same time, borders and boundaries are also important, as they provide a filter to all that may come rushing in: I must limit this force of the outside, monitor and shape its passage. I construct a set of parameters, however limited or tenuous, by which this body is made vulnerable. I give order to the intensities of which I am a part. According to a model of vibrant matters, subjectivity is reliant upon and made susceptible to its exposure to alterity.

Jacques Attali, in his extensive analysis of noise, provides a dynamic understanding of alterity, which can be useful in considering questions of overhearing and social exchange. According to Attali, noise functions as a force of interruption, one that announces, through its unmistakable agitation, a violence that always takes its aim at the social order. In response, as Attali suggests, systems of law and control are at pains to monitor, capture, and manipulate the intrusiveness of noise. "A concern for maintaining tonalism, the primacy of melody, a distrust of new languages, codes, or instruments, a refusal of the abnormal" – these features are common to systems of law and control.[12]

Noise, in this regard, is the force of the marginal and the different; a strange sound from a strange body which threatens the social order. Yet, noise is related to an inherent violence that fundamentally *underpins* the social order, as a primary thrust of audible intensity through which marginalities are mobilized and the expressivity of new languages intervene. Their intervention delivers into the social order a force of desire and festivity, usurping the normative patterns and tonalities by which society performs. In response, the social order adjusts, capturing the rupture noise delivers through mechanisms and rituals of production and consumption, which ultimately co-opt and regulate. Within such a system, music performs an essential role. For Attali, music captures noise within a particular ordering principle, giving it shape and form that locates noise back into the social order, into mechanisms of cultural production and consumption. Yet, music is prone to erupt, degrade, and mutate – to deform – under noise's continual pressure.

A type of negotiation surrounds the relation between noise and structures of power, one that is equally useful in considering relations between people. Questions of noise open onto experiences of intrusion and interruption, which I'm keen to suggest inaugurate new social encounters and relations. By doing so, the vitality and vibrancy of bodies may be seen as reliant upon the strangeness of others, as well as the basis for new forms of technological capture.

Richard Sennett, in *The Uses of Disorder*, develops an important view onto questions of interruption by applying the notion of "disorder" as a productive tool for nurturing social life and the diverse experiences intrinsic to it. Cities, as potentially dynamic spaces of diversity, provide a rich ground for deepening relations across demographic borders and ethnic divisions. For Sennett, "the great promise of city life is a new kind of confusion possible within its borders, an anarchy that will not destroy men, but make them richer and more mature."[13] From such a perspective, the author maps out methods for enabling conditions of urban intensity, an *anarchy* based on a belief in the human capacity to grow through social encounters of difference and uncertainty.

For example, Sennett suggests an approach to urban planning based more on "parts" than on the construction of a "whole form." As he states: "What is needed is to create cities where people are forced to confront each other so as to reconstitute public power ... The city must then be conceived as a social order of parts without a coherent, controllable whole form."[14]

Through an anarchic structure Sennett stages an extremely provocative inversion. While urban planners and social organizers may draw upon notions of cohesion and familiarity in establishing community, Sennett in contrast sees disorder as productive for deepening relations. Disorder may function as a means for encouraging social encounters, and for nurturing relations to what is unexpected or out of place. Disorder acts as a force of interruption, and in this regard, Sennett stakes out a productive understanding of the overheard – what I would emphasize as the potential of the fragment. A city of pieces. I hear something over my shoulder, next to me – a word, a sudden flow of dialogue; it cuts in: *what did they say?* It turns me, the intensity of this fragment, the power of the incomplete and even the incoherent. The overheard, following Sennett, is driven by a force of interruption, by what I may hear without truly knowing for sure: a fragment, and even a shard – *that pierces.*

The productive forcefulness of the overheard is one that exposes us within a scene of alterity: a figure, a voice, a sudden assemblage by which subjectivity is interrupted or unsettled. Through disorder, Sennett searches for these moments, prolonging them into a theory of urban planning as well as the basis for greater social tolerance. In doing so, cities become not so much spatial arrangements of streets and buildings, architectures and landscapes; rather, they operate as intensive zones of overlapping territories populated by a range of bodies and subjects, people and imaginaries. In short, Sennett's urbanism is a rough space of encounters whose lessons are greatly determined by the promises and challenges of what it means to live and work side by side with strangers.

In tandem with Sennett's urban proposals, the early theorizations on the modern city posed by Georg Simmel provide additional understanding on how disorder and noise contribute to the social life of the city. Simmel, in particular, brings attention to the figure of the stranger as one arising within the metropolitan scene; the stranger is explicitly one that stands outside particular social circles, that is peripheral, and yet at the same time, is proximate. As Simmel highlights, "[the stranger] is fixed within a certain spatial circle ... but his position within it is fundamentally affected by the fact that he does not belong in it initially and that he brings qualities into it that are not, and cannot be, indigenous to it."[15] The stranger comes to supplement and tense the intimacies and rituals of local life with *foreignness.*

The question of foreignness, and the trespass of the stranger, should not be posed in such a way that overlooks or neglects the ways in which territorial boundaries and ethnic divisions are deeply embedded within urban life; it may in fact matter greatly who behaves strangely, or occupies the position of a stranger within moments of disordering interruption, for such cuts and breaks may also incite abuse or criminalization. To be of a particular ethnic community always carries significance when trespassing territorial lines and easing close to social circles clearly of another order. This particular social tension may be placed as a horizon to the disordering possibilities posed by Sennett, a horizon that demarcates the inherent tension and violence embedded in any experience of rupture and exposure – to be caught off-guard, confronted with this figure I do not know or understand, is to enter a potential space of conflict. What Sennett intends is to use such encounters and uncertainties as the process by which to overcome assumptions and prejudices that may impose themselves onto the conditions of recognition and acceptance. In short, through prolonged exposure to the differences others present we may grow accustomed to and even enriched by what is other to ourselves.

Simmel sets out to map the particular social experiences instigated by this foreignness that the stranger exemplifies. Rather than understand estrangement as countering forms of being together, of community or cohesion, Simmel identifies it as participating within the growth of relationships, where "factors of repulsion and distance work to create a form of being together, a form of union."[16] Here, the stranger generates a relational dimension by intervening with the distance of the outsider, and by bringing peripheries or backgrounds into the inner life of the local. In short, the stranger traces around the known with what is not yet apparent or even conceivable, wrapping social relations with degrees of interruption and the continual introduction of remoteness, or what Simmel terms "objectivity."[17] A figure that looms in the background and that may step forward, or overstep unknowingly, to activate social encounters, as well as conversation and community.

May we understand the noisy interferences of the stranger as a discordant opportunity, one that gives way to new social relations? Following Sennett, is not the irritating force of interruption at times delivering explicitly what we might not fully understand? A noise that may challenge the tonalities of social community, but in doing so may equally enrich the vitality of its shape and form? As Simmel suggests, a voice that interrupts through its strange soundings may bring into relief the dynamics of new urban conditions. The interruptions and relations inaugurated by the stranger may incite a withdrawal into preserving ethnic or social lines; they may terrorize one into fear and trembling, yet they may also draw out an enriching encounter according to relational differences.

As a figure that cuts in, the stranger is a vibrancy that may disturb, but in doing so may also unwittingly fulfill our desires for cultural diversity – for what we may secretly desire: the wish to be disturbed, that is, to love and be loved: a love at first hearing. *This other that pierces.* The stranger, the overheard, the foreign, and the background – these are figurations and spatialities whose relational intensities rupture and extend processes of listening and of being heard – that *unhome* us – giving promise to what we may discover from the unfamiliar.

Following Jacques Attali's argumentation, which locates noise within a greater social operation where the possibility of cohesion is explicitly bound to primary forces of violence, of noisy confrontation and differentiating intensity, I'm led to emphasize noise as a property of (urban) relationships, and by extension, an arena not only of musical production and consumption, but also of social encounters that explicitly supports, through its disordering potentiality, a dynamics of alterity and social tolerance. If, as Sennett suggests, the anarchic potential of disorder may be applied to social spaces in order to develop the psychological skills for tolerating "painful ambiguity and uncertainty," overhearing the noise of strangers becomes prominent encounters within such anarchy.

In tandem with Sennett's notion of disorder as a productive break onto the urban fabric, I want to pose another form of interruption, one that equally supports encounters with alterity, yet through a shift in how property is managed. It seems, from the perspective of anarchic structures and forces of vibrant living, rights to the city are essential; struggles for housing, and for access to city life, do much to contour Sennett's theories with a political economy of space. Disordering principles, it would also seem, can only be nurtured by allowing for a diversity of housing and zoning laws, which enable people from a diversity of backgrounds and economic levels to inhabit the same territories. In this regard, we may consider the activities and culture of squatting, which have at their core an anarchic principle, and often carry into the political economy of urban space the desires for diverse living. Abandoned or uninhabited buildings, whose owners often use them as tax havens and repositories of income, are occupied and lived in by those who not only need housing, but often aspire to a realization of the urban vision Sennett presents, that of intensifying relations amongst the urban populace. Within cultures of squatting, anarchy functions not only through principles of self-organization, but equally as a sensibility toward broader relationships, especially against the capitalistic mechanism of profit and debt.[18]

Jamie Heckert, in an extremely thoughtful essay, develops a perspective on anarchy that may support understandings of alterity as grounded in care and compassion, as well as listening. From such a position, anarchism is offered as a positive affirmation of an "ethics of direct relationships," one that enriches processes of individual empowerment and collective organizing. "Nurturing autonomy in communities, workplaces, ecosystems and homes, it seems to me, necessarily involves getting on with others who experience the world differently."[19] As an ethics of direct relationships, anarchism may enable such a process, giving the needed skills and resources by which to co-operate. Specifically, for Heckert, this is based on conditions that support what he terms "listening, caring, and becoming," all of which form a strong basis for a commitment to "people being involved as directly as possible in making the decisions that affect their lives."[20]

Listening, caring, and becoming, these are essential principles of the anarchism that Heckert seeks to nurture, one that ultimately, like Sennett, aims to shift the conditions by which people join together to cohabitate and also, to work out how such cohabitation may support a more egalitarian living. As Heckert states, "rather than relying on fixed structures and rigid thinking, anarchism perhaps then involves developing a comfort with uncertainty."[21]

Practices of squatting are also often based on principles and ethics of direct relationships; people coming together around a form of autonomous action, which necessarily involves a process of self-organization, the sharing of differences, the working out of responsibilities, and the nurturing of a space for shared inhabitation. These coalesce into a culture of direct relationships, which radically alters the nature of property rights and urban behavior. Yet, this is often riddled with internal fighting, disagreement, fragmentation, and disappointment; direct relationships are fundamentally processes of struggle and the working through of differences. In this sense, they are vibrant matters assembling together multiple vectors of desire and opinion, as well as gleaned building materials, found opportunities, and shared knowledges and skills; they are reliant upon a strained commitment, one that places participants on the edge of criminality. This results in frictions with neighbors and local police, as well as solidarity and community celebration. Squats occupy this tense position, giving expression to the anarchic principles of self-organization and the disordered agitations Sennett seeks.

The occupation of sites and buildings may produce a type of break onto the governing patterns of spatial use and value; by repurposing a derelict or abandoned building, forms of squatting show us alternatives to how spaces can contribute to a larger sense of community, shifting value from built form to participatory experiences. For instance, the project of Prinzessinnengarten, in the neighborhood of Kreuzberg, Berlin, while not officially a type of squat nonetheless "smuggles" into the urban terrain an expression of self-organized and community-based space that subsequently acts as a model for alternative usages of urban sites as well as giving way to larger discussions on the value of public land. Urban gardening, bio-diversity, ecological rights, and relations between rural and urban communities, these come forward as civic practices as well as debates embedded in the site itself. The value of such initiatives and struggles is found precisely in their drive toward imagining other formations, and prompting critical debate on rights to the city.[22] They expose us to each other in new ways, nurturing the possibilities of not only meeting strangers, but also finding oneself sharing space and arguing about its future.

Urban disorders, and an ethics of direct relationships, work to give expression to the model of vibrant matters Bennett outlines by also locating it within the contemporary urban scene. The interdependencies and cohabitations inherent to distributed agency become a platform, an ethos, from which communities relate to strangers, and where the tonalities central to listening and being heard are reliant upon their own disruption. Noises are, in fact, expressive of disordering principles and resulting encounters; they are frictions that produce interruptions onto the plane of audition, scratching the surfaces and agitating the depths. In this regard, what we hear is deeply enriched by the intrusiveness and punctuations of what we overhear.

Networks / The Cognitive Body / Leaks and Invisible Remainders

Within this field of squatting and disordering relations, of vibrant bodies, it is important to integrate an understanding of the overheard also based on forms of surveillance: that the intensities of the stranger in the city may also be expressed through eavesdropping and other forms of spying. To overhear may be to take note of questionable speeches, the flows of gossip and secret information, to monitor and inform on one's neighbors. In the thick of overhearing, one is always made suspicious. Social encounters are thus not only productive of urban experiences, but equally perform to reveal hidden relations and to capture dissident talk. One may need to be careful *not* to be overheard – to disassemble. In the oscillations between strangers and friends, within the distributed vibrancy of restless bodies, secret agents are always active, gathering information and throwing suspicion into every exchange.

The means by which to instate systems of surveillance clearly gains in intensity and potential through the emergence of digital technologies. As the activities of Edward Snowden, Chelsea Manning, and WikiLeaks have dramatically shown, systems and technologies of covert intelligence wield great power in today's global environment.

Marshall McLuhan's analysis of the "electronic age" in the early 1960s is prescient of the conditions of global subjectivity that shape contemporary experiences. As he states: "In this electric age we see ourselves being translated more and more into a form of information, moving toward the technological extension of consciousness."[23] The extendedness of consciousness occurring due to electronic networking and automation produces another bodily configuration and sensibility from which we become intensely aware of the lives of others, and from which a "decentralized and inclusive" social formation takes shape.[24] Sensing beyond one's physical body, reaching outward and being reached, becoming involved in a network of relations – such are the conditions which, for McLuhan, are suggestive of a radical shift toward a "single form of consciousness."[25]

What is striking about McLuhan's analysis is the degree to which electricity and the electronic unfold the body and consciousness toward an externalization of sensuousness, feeling, knowing, and sharing. His ultimate proposition that we now "wear all mankind as our skin"[26] sets the stage for a complex transformation, one that suggests a sensitizing of the global situation within which individual experience is always already implicated and determined by a vast web of alliances and interferences, ruptures and reparations, *nerves and currents*, leading to what McLuhan further calls "retribalization."[27]

The topic of electrical vibrations finds parallel in Bennett's "vibrant matter" and Rolnik's "vibratile body" in which bodies and expressions of subjectivity are understood as "life forces" that are less about singularities and appear more as intensities of assembly, of matter-energy linkages; as we are *translated* into informational currents, *unhomed*, other formations are made possible (and demanded). The extended nervous system of McLuhan is a type of *mise-en-scène* that has become fully occupied by contemporary expressions and figurations of vibrant matters and unhomed subjectivities.

Franco Bifo Berardi, in his critical appraisals of the conditions of contemporary immaterial labor, positions McLuhan's rather optimistic global view within a space of exhaustion, paranoia, and loss.[28] For Berardi, the ways in which our lives are enmeshed in a vague yet persistent web of relations, that move beyond the skin and toward not only the nervous system, but toward what Berardi terms "the soul," functions to encourage if not incite a self-surrendering to any number of labors: while digital media give way to a range of individuated expressions, supporting matter-energy assemblages and vectors of desire and power, they in turn drive us toward a dizzying array of laborious operations: *a nervous shake.*

We never stop working or connecting; there is no end to being involved. The decentralizing of relations, through electrical flows and networks, may perform to support sensations of being a free subject while forcefully capturing one within a biopolitical structure of vital signals and controls.

Wearing mankind as our skin may form the basis for any number of ideological or technocratic performances, where one's nerves – *my translated subjectivity* – are made to serve a mechanism of production and consumption, passing along through its electrical pulses the operations of a purely capitalistic function. The "soul at work" never rests, allowing for potential alliances, sudden friendships, as well as the robotic operations that align us within a structure of tracking and transaction. An operational life in which one willingly performs.

With the nervous system extended beyond one's physical body, as McLuhan suggests, one is captured in a pressure of relations: a density through which my senses, my entire being, must navigate. As contemporary urbanisms expand and deepen, relating the physical body to an environment of intensified immediacy, they equally augment with the virtualities of digital networks, diffusing the body as an array of sensate and cellular signals: the nerves tremble and the animate properties that pass between oneself and a multitude of signals (re)distribute the body; the retribalizing enacted through the decentralized flows of the electronic brings one into a density of not so much figures or individuals, or even mass movements, but rather of "nano-operations," that is, the affective restlessness Bennett describes as central to contemporary life – the soul at work in today's unhomed condition of the perennially overheard.

Tiziana Terranova gives an insightful account of network culture by emphasizing how digital media produce mass culture and social grouping as well as intense segmentation.[29] This capacity to extend toward a global reach while supporting any number of micro-communities and digital solidarities, for Terranova, is reliant upon a dominating relation to images and their ceaseless circulation. Images act as "bioweapons" not so much delivering particular meanings, but functioning to produce affective intensities – "vibrations" as she calls them – around which specific cultural and political formations and identifications are mobilized or harnessed.

Following Jean Baudrillard's earlier analysis, in which representations (in the age of the spectacle) become untethered from grounded cultural meanings and instead operate as "simulations" or surface effects, the media flows of images continually charge the vitality of social relations with a force of restlessness.[30] In short, images not only pop up, they intervene directly into people's lives to shape daily routines, triggering responses and galvanizing people; they work to charge the political economy of attention and the social groupings to which network culture is conducive.

The circulation of image-flows delivers not so much particular meanings as forceful interruptions that ripple across the senses to assemble together the nerves of so many bodies, acting upon the world as "vibrations"; in this way, they are, in effect, agents and actants, *life-forces.* Following the "vibrant" model of Bennett, image-flows produce not only semiotic relations, or visual cultures, they in turn affect the rhythms of breath and nerves, and the restless linking of network assemblages. As Terranova enigmatically suggests, there exists an "invisible remainder" to the media flows of images, a remainder hovering around a particular image and by which sensibilities and cognitive labors are nurtured and directed.

What is this "invisible remainder" exactly, this addition that hovers in and around specific visualities? And to which the body is made responsive? I want to return to McLuhan, and in particular his additional proposal that electronic media locate us within an oral, and by extension, acoustical condition. As he states, "Because of its action in extending our central nervous system, electronic technology seems to favor the inclusive and participational spoken word over the specialist written word."[31] Such a statement is based on a general demarcation between the written and spoken word, and by extension, between the ocular and the auditory. For McLuhan, literacy produces a subject defined by looking, and the operations of an ocular knowledge, which steers us toward rational objectivity, a *looking upon* the world; a semiotics of visual meanings from which cultures are produced and shaped. In contrast, a society or community dominated by orality tends toward a process of interaction, *entanglement,* where listening generates an intensity of relations that requires continual negotiation; in short, while the written word allows for a certain detachment, a capacity for "separateness, continuity and uniformity," a linearity, the spoken heightens "pluralism and uniqueness," and even "discontinuity" – in short, active sociality.[32] It is this pluralism, this discontinuity and intense sociality that McLuhan identifies in network culture, leading to a conceptualization of the electronic age as one of orality, as being defined by an acoustical condition. This can be glimpsed in what Terranova highlights as the pervading dynamic to network culture, that of "intense segmentation," leading her to characterize network culture as a "permanent battlefield" between mass populations and intense localities. Network culture, in other words, is a "space that is *common*, without being *homogeneous* or even *equal*."[33] A space, as McLuhan suggests, of "pluralism and discontinuity" – active sociality.

The intensity of involvement – of vibrancy and vitality, of matter-energy assemblages – produced by network culture may be understood to displace the capacity of the written word, as appearing from a position of continuity and uniformity, as a stable signifier. In contrast, the ceaseless flows of expressions and articulations, comments and commentaries, sharings and postings – *translations* – embedded within network culture operate precisely in support of a pluralism – a *restlessness*; the soul at work to which the spoken and the acoustical are expressive. Within this field of vibrations, of network assemblages, one's expressivity – of mediated productions – contributes to the functioning of nano-operations, of subjectivities endlessly fragmenting and regrouping, whose nervous systems, cognitive labors, and cellular bodies perform connections.

Subsequently, the animate and the sensate intensities of network culture may stabilize into sudden apparitions, *images*, yet they do so through a force that is never stable or purely imagistic. Images, instead, are immediately connective, affective assemblages across matter and energy. As Terranova suggests, they prod and poke through a vitality that is more vibration than image, more oscillatory than pictorial, to set one's body on edge and to deepen the drive of network relations.

McLuhan's acoustical descriptions support this questioning, suggesting instead that images take on the conditions of an electrical oscillation, a sonic intensity, as a pressure of the molecular. The remainder that Terranova identifies, invisibly at work in support of image-flows, is precisely that animating periphery that pulls at our senses, that is already closer than imagined, and that produces a political economy of attention and mediation through its ability to generate an overflow of desire and involvement, to capitalize upon one's affective and restless activeness.

The remainder is nothing other than oscillations and currents - *agents* - trafficking the capitalistic flows of input/output, trembling with the temporality of the appearance/disappearance instant, a propagation of vibratory excess. *Something I know already awaits me, that overhears: the image that unhomes.*

Returning to the theme of the acoustical, and the question of image-flows, we might better understand networks and their movements as *molecular agitations* that assemble subjects/objects while at the same time diffusing them, holding them in an anticipatory net of nano-operations. By extension, these vibratory oscillations and excesses must also include an array of "listening" agents or devices, robotic "tentacles" by which to pick up and police the assemblages of matter-energy and their potential. In short, the network is one conditioned by *overhearing*, as that which constantly delivers the background to the fore, that presses its forceful interruption onto the senses by delivering strangers and strangeness into one's attention, and that defines subjectivity, this body, as a cognitive agent - adept at intervention, active restlessness, and social production. The invisible remainder that Terranova speaks of is suggestive for this state of overhearing I'm engaging, in which identity is defined by the processes of network culture, prodded and distributed by any number of affective and generative vibrations.

Unsitely Commons / Pirate Cultures / New Moral Challenge

Vibrations and vitality, unhomed subjectivities and the informational pressures and promises of network life - these are coordinates by which contemporary listening is situated. And according to which we search, through the flows of a "nomadism of desire," for each other, amidst crowds that are both local and distant, imagined and sensed, on the city street and equally on-line. I am at once a friend and stranger within this new spirit of biocapitalism, always already at work, ceaselessly oscillating between languages and translations, connections and counter-connections. Under such conditions, one also searches for tactics, a set of acts and positions through which to ground oneself, and to give commonality to the perpetual exposure to alterity that is the basis of global culture. In short, to negotiate the interruptions that pervade the experiences of contemporary life.

In such a new state, the operations of network movements continuously introduce us to a plethora of currents, bodies, voices – *expressivities* – fragmented and flashing across one's extended perceptual field. In doing so, network culture shifts from the imagistic to the affective, from the "alphabetical" to the "acoustical" (to use McLuhan's vocabulary), animating, through hyper-connectivity, the territories of the not-yet-apparent hovering in the background – *this background that is always shifting*, that is closer than imagined. The relationships we develop through such dynamics and animations – such vibrancy – find recourse through understandings of the stranger and estrangement; proximate and distant, near and far at the same instant, the contemporary subject becomes adept at negotiating and utilizing the unsteady and difficult murmurings of a strangeness so near. Here, I am relating less to what is explicitly in front of me, and instead, am always engaged in relations with what is surrounding, perennially to the side of my attention – that is, through a disordering principle in which social encounters are charged by constant strangeness, to those whom I overhear and am overheard by. This anarchic condition, extended from Sennett's urban theories and McLuhan's conceptualizations, ultimately redraws how we understand the sphere of public life, and an ethics of direct relationships, to force other dynamics into acts of relating – of speaking and listening to each other.

Networks may be understood as being dramatically conditioned by what Michael Warner, in a discussion on the question of the public sphere, terms "stranger sociability."[34] As Warner highlights, the public sphere is based on sustaining particular communities and their discourses, while at the same time addressing its concerns to "imagined others" – those that may sympathize or identify with a given articulation of concerns, or with a particular social group or community. In this way, publics are often fringed by strangers toward whom they partly direct themselves.

Warner's notion of a "stranger sociability" runs rampant within the network, and locates one's senses within a field of operations that is always delivering more than expected: attention is put under pressure, excited and prompted, capitalized upon, and accordingly, one is subsumed by an operational logic – a "techno-semiotic" functionality through which we come to relate. In short, I am myself a stranger and I enact my strangerhood in ways that always creep up on others: I poke you, I comment,

I *unhome* you, I am always eavesdropping, waiting for opportunities to intervene, to overhear, as well as to assemble or collaborate.

Network sociability is a noisy production of *overhearings* whose figures and voices are proximate and distant at the same time. From such a position, public power is founded on being able to move between multiple polarities: between the virtual force of the network and the intensities of social contact, this stranger sociability. As Maria Miranda suggests in her work *Unsitely Aesthetics*, productions that draw network activities along with local physical sites into equal play construct a range of performative arenas by which new formations of critical and creative togetherness are defined.[35] For Miranda, these are based upon what she terms "uncertain practices," exemplified in artistic projects that work to negotiate the realities of "mediated public spaces." Shifting between local contexts and global networks, unsitely aesthetics generate new relations that intensify the power of a work as a form of social engagement by distributing and dispersing its creative force as a "new possibility for public encounter."[36] From complex stagings of hacktivistic actions to narratives and expressions that maneuver across the borders between on- and off-line conditions, Miranda identifies the arena of connection as one of new relational possibilities and critical inquiries. In this regard, uncertain practices draw out the unsteady and ambiguous, but no less palpable complexity of mediated life, thereby giving expression to the new ontology of the social. For these are not so much augmented realities as they are agitated connections that squat the new architectures of digital life.

The *unsitely* conditions that Miranda articulates in this era of mediation is suggestive for considering the politics at play in being connected, drawing us closer to the issue of "digital sovereignty." The relation between sited territories of local communities and the network systems that link us to global communications is one deeply shaped by geopolitical projects, corporate mechanisms, and governmental agencies. This is aptly demonstrated by considering the recent movements within Latin America, stemming from continual US interventions within the nations of the South, to develop independent media infrastructures. For example, President Evo Morales of Bolivia, in his call for united action against the imperialistic practices of the US in particular, has moved forward on building a digital system of network communications – a sovereign cloud – independent from international corporate and political interests and infrastructures that make national communications and populations susceptible.[37]

How are we to understand the call for digital sovereignty and its place within globalized conditions? To what degree can digital sovereignty resist particular colonization, and what new practices might it generate to assist in rethinking the notion of sovereignty today as well as what it means to connect?

Digital sovereignty requires an active tussle with the conditions of the vibratile, the overheard, and the estranged that are always challenging borders and which position us as *unhomed* subjects. As such, securing digital sovereignty is an attempt to contend with the powers that constitute contemporary realities so as to forge direct relationships across or against network life. Within the vibrations and vibrant matters that assemble one's cognitive labors and loves, new resistances become imperative in order to negotiate the political economies of attention and mediation, which traffic in both territorial colonization and cognitive capture.

Morales' call for digital sovereignty raises important questions as to the conditions of network culture and its relation to grounded political struggles. We may find an additional example by considering the movement of Indigenous Communications whose project underscores the complex relation between digital networks and their significance for local emancipatory struggles. While global culture provides the means for any number of transnational movements, from imperialistic capitalism to the art of commoning, it also places pressure on existing cultures, such as indigenous First Peoples worldwide, that struggle for sovereign rights. These communities search for ways to "shelter" and protect themselves from the encroachments that have always oppressed them and that intensify under new global networks. In this context, the movement of Indigenous Communications develops discourses and practices that resist or reshape the vibratile conditions of network culture, aiming to enhance new territorial securities.

Indigenous media rights are ultimately a struggle over the knowledges, practices, and languages inherent to specific communities that continue to experience colonization and control by dominant powers. In this regard, they remind us how satellites, optical fiber cables, cellular towers, and server farms, for example, carry specific corporate and governmental interests and economies that make "connecting" a question of geopolitics. The project of indigenous media is therefore deeply susceptible to the controlling infrastructures of digital systems and their codes, which shape the flows of knowledges, practices, and languages, and the possibility of sustaining community traditions.

In seeking to establish digital sovereignty might indigenous media struggles remind of the necessity to resist the smooth operations of network culture, calling to mind the territorial struggles embedded in the digital experience? Returning to the issue of disorder, and anarchic formations of self-organized culture, can we understand indigenous media struggles as acts of squatting that seek to claim a form of public power on their own terms? Occupations of networks by which to realize the promise of digital commons, as that which fights for ensuring plurality? Indigenous media struggles articulate the importance of the "digital commons," to ensure an "open space" for respecting diversity while protecting equal rights.

By *estranging* the languages and codes of media flows, occupying the airwaves through community broadcasts, threading indigenous narratives and knowledges into the stream of digital communications, and diversifying the field of network practices with their particular traditions, indigenous media reach beyond dominant structures to "offer an alternative model of grounded and increasingly global interconnectedness."[38] As such, indigenous media struggles remind us that in the seemingly free movements of network culture, capitalistic and imperialistic power is deeply at work, and the commons is ultimately a question of the practices by which network relations may enable resistances.

In this regard, indigenous media steer us toward a general framework related to pirate media and culture which operates through appropriations of existing energy grids and networks to manifest what Ravi Sundaram terms the "media infrastructures of the poor."[39] Pirate practices equally interrupt the (capitalistic) flows of media conglomerates, arguing for open-source access and creative commons, and a sustainable model of media autonomy, often from the perspective of the Global South. Yet, with the emergence of the Pirate Parties International, the issue of open media is housed within a greater political agenda including questions of territorial resources, national sovereignty, and global responsibility. Stemming from struggles over privacy and media access in Sweden and other parts of Europe, pirate parties worldwide begin to articulate a new form of democracy by linking across nation-states, and developing new structures of citizen participation, which is giving way to a new frontier of political governance.[40]

Patrick Burkart, in his study on pirate politics, highlights the expanded dimension at work in the pirate party movement, suggesting that, by instigating a "decolonization" process of the Internet, pirates form a movement that "gives notice" of a "fundamental clash between cultural codes and technocratic routines."[41] In this regard, pirates stage an important challenge to practices of global extraction and the enclosure of public resources, as well as to our institutions of democracy, seeking to enhance equal access and global responsibility.

The project of the Pirate Parties International can be understood as an attempt to counter the intensification of global disenfranchisement. Zygmunt Bauman examines our contemporary condition by way of three specific features at work in today's experience of "unhomed subjectivity," all of which suggest the need for greater political and moral responsibility. As he outlines, the reduction of policies and programs of social welfare and public spending throughout Western states has led to an intensification of contemporary "precarity."[42] At the same time, as states work to relate increasingly to global structures of economy and power, there is no equivalent level of "political enfranchisement" occurring on a global scale; instead, practices of citizenry remain locked within national structures that do less to nurture civil society while drawing upon and being susceptible to the flows and deregulated speculations of global capital. Finally, as Bauman suggests, the realities of expanding global relations "constrict" the needed development of a new moral sensibility: confrontations with the movements and migrations, the unhomed conditions of global experiences, ultimately *stress* our moral bearings, normalizing the intensifications of social conflict, war, and violence worldwide. From the rending experiences of contemporary life, shaped by media flows and related biocapitalistic economies, which construct a new ontology of social relations – of being with and learning from others – one must begin to reorient political behavior and moral responsibility.

The disordering and interruptive experiences of contemporary life requires a new sensitivity for others, as well as a new spirit and understanding for the complexity at work today: "wearing mankind" as a skin is ultimately a rending experience – a continual tearing apart of the social body – and while McLuhan defines the "global village" as a horizon of integration and multiplicity, the reality has proven far less equitable. In order to reorient political behavior and moral responsibility, the sensitivity for "sheltering" and "homing" those in need must work against or through the new state of global media and economic instruments that expel us from structures of security, intensifying the need for common sharing. In this context, indigenous media struggles demonstrate the susceptibility some communities dramatically experience under new global flows, which often undermine the dispossessed only further. Parallel to this, pirate culture, in its dedication to open media platforms and transnational alliances that shadow capitalistic movements with acts of hacking, sampling, and open networks, exemplifies the necessity to work through the unhomed state of network culture – to inhabit the volatile channels of contemporary life with a new sense for shared responsibility. From such positions, questions of sovereignty and the commons are brought forward: how to withstand the imperialistic tendencies of contemporary power that evict and expel while nurturing the vibrant assemblages that afford coalitional frameworks of resistance, commoning, and making together? To resist take-over while remaining open and aligned with others? To figure ways to ground ourselves through the conditions of global estrangement and disenfranchisement?

I'm interested to return to the figure of the overheard, which I've highlighted as the basis for listening today: that under the conditions of contemporary urban life, and according to the interruptiveness of network culture, one is situated within an ever-shifting set of coordinates, unsitely and vibrant, that require radically new orientation. As such, listening and being heard are shaped through a political economy of attention and mediation. Accordingly, biocapitalism functions less through material structures and more through nano-operations, which conduct a certain bio-cognitive performance according to one's willing participation in digital systems. Within such conditions, I've attempted to describe overhearing as a force of capture as well as potential alliance: from our unhomed subjectivity, this body always already defined by the logic of vibratility, which makes one available to an array of controls, one searches for friendships and collaborations, associations and coalitions. Overhearing and being overheard become new skills and new frameworks from which to gain resources, from which to align and assemble, and through which to also interrupt, as well as to contend with the interruptions that define one's articulations and social exchanges.

Might overhearing suggest a model of listening that by steering one through conditions of interruption enables gestures of caring beyond the familiar? Is it possible to consider the overheard as the basis for uncertain practices, those that may defend the new state of vibrancy as one aligned with moral orientation and responsibility? If the overheard turns us toward the other, might this support the figuring of new political subjectivity – a subject prone to what Felix Stalder outlines as "network individualism," which interweaves individual distinctiveness with collective sharing?[43]

It is in this sense that I understand what Kate Lacey terms "freedom of listening." While freedom of speech is often posed as a critical framework, Lacey instead turns us toward listening, for it is in listening that "an active, responsive attitude inheres" and from which a plurality of opinions, expressions, and voices are heard.[44] Freedom of listening is a call for giving resonance to voices often unsited and unhomed by the political economies of attention and mediation. As indigenous movements remind, public arenas and discourses are greatly shaped according to communicational flows and stoppages, of not only having access, but also of contouring and regulating that access with particular content free of the potential intrusiveness of those who control the code and related apparatuses. In short, to secure one's freedom of speech and listening through digital sovereignty and with respect to the digital commons.

It is my view that in overhearing one develops the needed skills for being attentive within this economy of attention; to enact one's freedom of listening in order to contend with the new ontology of the social, which is one of continual assemblage and interruption. In this regard, I'm drawn to consider how overhearing may support new practices of care: to hear and attend to what the stranger says. Within this movement one may overhear precisely what needs to be heard, and one may find the means for bringing such listening into the arena of public concern. Is this not what Rolnik and Bennett suggest by way of "matter-energy assemblages" and the new ecologies of contemporary experience, to find within the disasters of global capitalism the needed support structures and collaborations by which to enfranchise citizens to act as guardians of earthly life? That in the velocity of global hyper-connectivity, which is always serving the project of imperial gain, agentive positions may be nurtured to give shelter and to home the common earth, its indigenous cultures and its pirate politics. Overhearing may present opportunities for easing into others' lives with the desire to join in the increasingly complex struggles that affect all of us as global citizens.

Here, we may learn from the intensely fragmentary nature of one's unhomed subjectivity the potential of interruption, which, according to the dynamics of global culture, may paradoxically form the basis for *grounding* the struggles and ideals of direct relationships through a range of alliances and a new urgency for practices of listening, caring, becoming. Acts of interference and estrangement may assist in generating assemblages and linkages by which commons may arise precisely as practices of caring for diversity: to create new cultural structures for nurturing not only direct relationships, but the possibility of sheltering those relations. As Kwame Appiah suggests, it becomes a question of taking the "minds and hearts formed over the long millennia of living in local troops and equipping them with ideas and institutions that will allow us to live together as the global tribe that we have become."[45]

Notes

1. Jane Bennett, *Vibrant Matter: A Political Ecology of Things* (Durham, NC: Duke University Press, 2010), 54.
2. Ibid., 38.
3. Ibid., 57.
4. Suely Rolnik, "Anthropophagic Subjectivity," in *Arte Contemporânea Brasileira: Um e/entre Outro/s* (São Paulo: Fundação Bienal de São Paulo, 1998), 2.
5. Ibid.
6. Ibid., 10.
7. Homi K. Bhabha, *The Location of Culture* (London: Routledge, 1994), 15–17.
8. Rolnik, "Anthropophagic Subjectivity," 15.
9. See Judith Butler, *Precarious Life: The Powers of Mourning and Violence* (London: Verso, 2006).
10. Bennett, *Vibrant Matter*, 49.

11 Ibid.

12 Jacques Attali, *Noise: Political Economy of Music* (Minneapolis: University of Minnesota Press, 1988), 7.

13 Richard Sennett, *The Uses of Disorder: Personal Identity and City Life* (New Haven: Yale University Press, 2008), 108.

14 Ibid.

15 Georg Simmel, "The Stranger," in *On Individuality and Social Forms: Selected Writings* (Chicago: University of Chicago Press, 1971), 143.

16 Ibid., 144.

17 Ibid., 145.

18 For an informative history on squatting within Europe, see Bart van der Steen, Ask Katzeff, and Leendert van Hoogenhuijze (eds.), *The City is Ours: Squatting and Autonomous Movements in Europe from the 1970s to the Present* (Oakland, CA: PM Press, 2014).

19 Jamie Heckert, "Listening, Caring, Becoming: Anarchism As an Ethics of Direct Relationships," in Benjamin Franks and Matthew Wilson (eds.), *Anarchism and Moral Philosophy* (Basingstoke: Palgrave Macmillan, 2010), 190.

20 Ibid., 191.

21 Ibid., 193.

22 For more on Prinzessinnengarten, see the interview with founder Marco Clausen in the free newspaper, *Free Berlin*: http://errantbodies.org/Free_Berlin/FreeBerlinNo.3.pdf (accessed October 2016).

23 Marshall McLuhan, *Understanding Media: The Extensions of Man* (London: Routledge, 2002), 63.

24 Ibid., 8.

25 Ibid., 67.

26 Ibid., 52.

27 Ibid., 26.

28 See Franco Bifo Berardi, *The Soul at Work* (Los Angeles: Semiotext(e), 2009).

29 Tiziana Terranova, *Network Culture: Politics for the Information Age* (London: Pluto Press, 2004), 150.

30 For more on the issue of simulation, see Jean Baudrillard, *For a Critique of the Political Economy of the Sign* (Candor, NY: Telos Press Publishing, 1981).

31 McLuhan, *Understanding Media*, 89.

32 Ibid., 91–95.

33 Terranova, *Network Culture*, 154.

34 See Michael Warner, *Publics and Counterpublics* (New York: Zone Books, 2002).

35 See Maria Miranda, *Unsitely Aesthetics: Uncertain Practices in Contemporary Art* (Berlin: Errant Bodies Press, 2013).

36 Ibid., 14.

37 International Summit against Imperialism (2013), in which Morales announced a six-point strategy for sovereignty. See: https://libya360.wordpress.com/2013/09/27/bolivia-against-colonialism-and-imperialism-six-strategies-for-sovereignty-dignity-and-the-life-of-the-peoples/ (accessed December 2016).

38 Faye Ginsburg, "Rethinking the Digital Age," in Pamela Wilson and Michelle Stewart (eds.) *Global Indigenous Media: Cultures, Poetics, and Politics* (Durham: Duke University Press, 2008), 304.

39 See Ravi Sundaram, *Pirate Modernity: Delhi's Media Urbanism* (London: Routledge, 2010).
40 For more on the Pirate Parties International, see: https://pp-international.net.
41 Patrick Burkart, *Pirate Politics: The New Information Policy Contests* (Cambridge, MA: MIT Press, 2014), 126.
42 See Zygmunt Bauman, *Strangers at Our Door* (Cambridge: Polity Press, 2016).
43 Felix Stalder, *Digital Solidarity* (London: Mute Books, 2013), 25.
44 Kate Lacey, *Listening Publics: The Politics and Experience of Listening in the Media Age* (Cambridge: Polity Press, 2013), 177.
45 Kwame Appiah, quoted in Bauman, *Strangers at Our Door*, 72.

4

The Itinerant

Experiences of homelessness and eviction run throughout the economics of austerity prevalent in the wake of the 2008 banking crisis. In tandem with the general dismantling of social welfare systems across Europe and the UK, debates around social housing, for example, have led to ongoing demonstrations and protests against privatization and, by extension, the project of neoliberalism. Property and its loss become key referents in struggles for the right to the city and the right to affordable housing.

In tension with such realities, migrancy and the forced movements of people around the world, stemming in particular from conflicts in the Middle East and parts of Africa, have compounded debates and struggles in terms of not only housing and city life, but equally homeland and the right to remain or return. The No Border campaigns across Europe, initiated by activists and migrant movements, underscore the global reality of transnationalism – defined increasingly by what Saskia Sassen terms "predatory financial instruments"[1] – and which have additionally spurred anti-immigration movements and the fragmentation of European unity. Economic and border crises thus exacerbate existing struggles over housing and urban rights, giving way to an increasing presence of bodies out of place, on the streets and peripheries, without home or places of belonging, families and individuals stranded and in need of support. Transience, migrancy, homelessness, and the itinerant, while always being experiences of human reality, have taken on pronounced intensity and number within the contemporary global environment.

The sociologist AbdouMaliq Simone suggests that urbanism is fundamentally based upon the circulation of bodies, materials, and goods; that it facilitates and produces an array of movements from which entire power structures, of capital and political dominance in particular, are supported.[2] Additionally, urban conditions afford the movements of bodies – urban subjects are always gaining (and losing) power and opportunity from structures of circulation, finding solidarity and creating coalitional frameworks, as well as everyday relationships, through possibilities of urban movement. Yet, at the same time, such movements and relationships pose a risk to the maintenance of governing power structures, as well as to the ongoing productivity by which such structures are sustained. In short, urban subjects may overflow, occupy, interrupt or redirect the circulatory flows and infrastructural conditions of cities at any given moment. Hence, Simone observes, urbanism requires a type of policing, a *settling* of bodies, and a stabilizing of the potent flows inherent to urban culture. Bodies must equally be grounded. As can be seen in various laws that make vagrancy or loitering illegal, being itinerant or homeless (or aimless) are embodied expressions that challenge, by standing out or falling off the grid of city structures, the stabilizing force necessary for urban productivity.

Produced by current economic conditions, while being subjected to any number of discriminations, transience and being without place are pronounced consequences of a neoliberal system, for neoliberalism forces an intensity of movement – of mobility and hyper-connectivity, of insecurity – which creates conditions of possibility while making home and place vulnerable to the powers of economic gain and loss. How fast may one move in order to secure a livelihood in today's urban assemblages of finance and power? How might one's mobility lead to sustained relations within communities, or between friends and families? How do the hyper-connective operations and general itinerancy necessary for neoliberal survival shape my sense of place and acts of place-making, especially when the distinctions between home and work, friends and colleagues, become blurred under the pulse of post-Fordism?

On one hand, I am mobilized by the possibilities of being involved in multiple places and, at the same time, I am pressured to sustain relationships emerging from such conditions, most of which continually shift between professional and personal languages, locating intimacy and deep community in relation to entrepreneurial work. In short, displacement is central to neoliberal productivity, one that invigorates one's psychic and emotional life with feelings for possibility – to extend oneself into all sorts of relations – while bracketing them within a general precariousness in which mobility paradoxically locates one upon a threshold to homelessness. Eviction and expulsion have become not only experiences related to losing home due to burdens of personal debt, but rather a "new logic" governing relations between people and place, governments and planetary resources, political economies and global transience.[3]

Against the prevailing logic of expulsion, gestures of occupying, standing still, slowness, laziness, and commoning have gained traction to galvanize individuals around concerns for collective sharing that lead to new forms of creative and critical togetherness. These gestures become tactics aimed at shifting the pace and intensity of neoliberal economics and behavior. In contrast to the drive of competition and privatization, and against the ethos of entrepreneurial culture and hyper-connective exchange, which partly transforms conversations into transactions and the deep meanings of voice and speech into productions of information, people are moving toward other modalities, in search of what Franco Bifo Berardi terms "pockets of friendship" and "nodes of resistance."[4]

I would highlight these gestures and actions as forms of critical and creative exit, where people transpose the circulations and mobile intensities of contemporary global urbanism onto new formations of being together for the sake of shared resistances and independent initiatives. Rather than understand circulation and movement as serving the gains of capitalistic mechanisms, possibilities for enriching the dimensions of intimate and collective relations, of sharing and deepening life with others, across and through the precarious conditions many inhabit today, are enhanced and nurtured – mobility comes to also serve a *slowing down*; not so much a spending of time, rather a giving of time by which "precariousness and sadness can become something different."[5]

Transience is therefore not only framed by experiences of neoliberal pressure and entrepreneurial opportunity, but something that may contribute to counter-narratives and what Stavros Stavrides terms "communities in movement,"[6] which take hold of their precariousness, their local connectivity and mobility, along with the dialogical support structures born from network culture, and extend them into initiatives of critical and creative togetherness – communities that attempt to exit the conditions of the neoliberal city in order to collect resources, making them common to the cause of alternative instituting. Experimental pedagogical structures, alternative economic models, food banks, urban garden projects and neighborhood assemblies, artist collectives, hacker labs and open networks, these come to dominate the culture of communities in movement – unlikely publics that reorder the logic of the public sphere today. These movements and publics come to produce what Nick Couldry terms "intensities of listening" which reshape the drive of a neoliberal logic by insisting on the importance of sharing, voicing, and telling.[7] Within backrooms in Whittier, California, empty warehouses in Barcelona, or urban lots in Athens, these communities and intensities may do much to construct significant cultures of social solidarity against normative economic structures.

I understand these cultures and intensities of listening as "lyrical" expressions in which small-scale instituting initiatives and temporary gatherings strive for a new common life, and are often directed by creative principles of radical sharing. In taking charge of the productions of mobility and cognitive currencies at work in neoliberal globalism, unlikely publics forge new sensibilities based upon what Michael Warner terms "poetic world-making."[8] In his work on counter-publics, Warner suggests that "public discourse" is often misunderstood as being based solely upon the gathering of "already co-present interlocutors," or through the circulation of a "prevailing" language to which others adhere.[9] In contrast, Warner suggests that publics and their discourses are forged through a "performative" operation, one that must point toward a particular world-view while constructing that world through its speeches, vocabularies, idioms, spatial and material forms, rituals, shared joys and shared indignation, etc. Importantly, public discourse is never only about "understanding" and "orientation" within an existing framework or environment. The languages and actions that come to nurture public life are equally based on the crafting of an imaginative field – an imaginary itself – which people move to inhabit and give life to.

Poetic world-making, in the context I'm mapping, is emphasized as a critical demarcation against the ways in which creativity, in particular, has been incorporated into a neoliberal agenda: creative cities and creative economies as projects align themselves with "the creative class" and gain traction within an environment defined by the mobility and hyper-connectivity of global culture. In contrast to these projects, the labors of communities in movement and unlikely publics that result in the slow construction of alternative instituting, that gather the knowledges and desires each person carries into soft structures of critical and creative togetherness, and that work and overwork themselves without much or any economic support for the benefit of a collective intelligence and imagination, these labors underpin a fragile yet persistent, and extremely resourceful and determined, lyricism: narratives and oralities, soundings and intensities of listening that manifest a faith in and affection for each other as people invested in a responsibility for the world.

Franco Bifo Berardi further captures the force of this cultural lyricism by underscoring "the poetic" as a way of being and speaking. In particular, the poetic is emphasized as a means for retrieval and recuperation of language from out of the "techno-semiotic linguistics" of neoliberalism, one that may assist in constructing "a social sphere of singular vibrations intermingling and projecting a new space for sharing, producing, and living."[10]

If experiences of precariousness and expulsion often incite highly conservative, racist, and ethnically divisive discourses and social movements, how might we understand and enhance the "vibrations" of global itinerancy as poetic weapons against conditions of eviction and discrimination? Can resistances found from mobility and hyper-connectivity lead to new states of political imagination, where transience and the itinerant become a ground for new formations aimed at equal care as well as making home?

Echo Worlds / And Diversal Subjects / Lyrics of Displacement

The bodies that rush by and that send vibrations across the floor, and that move like bundles of energy suddenly let loose, surging and flowing around each other, and whose voices puncture the air with their languages – *sound is movement,* and the voice is essential to extending oneself into the world: one speaks, one vocalizes, and this sound moves from oneself into spaces and toward others, to nurture relations, and to announce, through such gestures, that *I am here.* And further, that I hear you.

It is important to highlight that although I attempt to speak about sound and its expressivity, to capture this sound here and to learn from its behaviors, I am speaking of something that has gone missing. In short, I am always speaking *after* sound, running behind it, even as it reoccurs continuously (*it is always beside and beyond me, this sound, this voice that I hear*). The fleeting and punctuated event of sound is one of transience and transition; an itinerant and migratory sensorial matter, sound is both a thing of the past and a signal of the future; it points us toward what has happened – for every sound is an index of an event that, by the time we hear it, has already transpired – while equally pulling us forward by echoing beyond, toward a distance over there. The articulated presence of any sound, at one and the same moment, is to be found in its disappearance and its becoming.

What might itinerancy provide or suggest for conditions and experiences of listening? Is there a form of cultural production, a social act, or a subject position that may be supported by the fleeting and evanescent qualities of sound and sounded action? A sense for being a subject in relation to others, a necessarily ephemeral condition and from which one learns to articulate against the odds? Do such migratory behaviors give guidance for fathoming what has occurred and what may still take place, in a future to come? And for evoking, through any number of gestures, the supportive structures or matters that may aid in times of transition, loss, and loneliness?

The transient and the itinerant, the migratory and the expelled, these are modalities by which bodies move, conditions by which each journeys and struggles. To shift localities, to transition or transfer, movements and experiences that deeply align physical location with psychic life – to *take flight* is fundamentally an act of psychic labor, punctuated with dreaming and loss, a traumatic fragmentation of identity and identification that requires deep emotional resources: to compose oneself around new sensations, perceptions, and languages as well as social structures and local relations. One figures a way, and is in turn figured by all that one confronts as a foreigner; a type of orchestration riddled with noise and rupture, sudden links and resonances, offered assistance and conflicts of interest. To move or migrate entails an entire range of negotiations, and, accordingly, one must shift the register of the voice, of this speech that needs to enter conversation, to be understood, while also finding the confidence to ask or speak back. Physical locations and psychic labors, along with the markets and conflicts that often incite bodies to move and migrate, or to flee, form a constellation of forces, underscoring mobility equally as a question of political conflict and economic capability, as well as the affective attunements between place and people by which to survive.

To deepen an understanding of the transient, and what it might lend to emancipatory practices, I want to turn to the post-colonial work of Édouard Glissant.[11] Drawing from Caribbean literary culture, Glissant seeks to elaborate a theoretical framework by which to think through and beyond colonial histories and subjects. In particular, questions around the complexity embedded within histories of enslavement, additionally shaped by the cultures and topographies of the islands, lead Glissant to a range of compelling ideas. For instance, his theoretical activity develops out of an engagement with Creole languages and culture (*creolité*), which are mobilized as the basis for a greater field of dynamic thought. Creole languages, which integrate and mix (at least in the context of the French Antilles) French, African, and indigenous Caribbean languages into new linguistic forms, come to provide the basis for reshaping the colonial framework. In short, such processes of intermixing are understood and elaborated through a notion of "creolization." "Creolization is about the mixture and continuing admixture of peoples, languages, and cultures. When creolization occurs, participants select particular elements from incoming or inherited cultures, endow them with meanings different from those they possessed in the original culture and then creatively merge them to create totally new varieties that supersede the prior forms."[12] In this regard, creolization is a complex manifestation of hybridity and intercultural confrontation – what Jean Bernabé, Patrick Chamoiseau, and Raphaël Confiant in their seminal essay "In Praise of Creoleness" would term "Diversality." "Our history is a braid of histories," they write, pointing to the complicated ways in which local cultures are always the result of an interweaving of multiple voices, a diversity that is equally a trajectory of cultural intensity.[13] A braid, a *diversality* constructed and specifically linked to the diaspora of the Caribbean: the scattering of people from their homeland which, on the islands of the Antilles, forms into "the world diffracted but recomposed."[14] The world as it comes to the shores of the islands, driven by the European enslavement and transport of Africans (as well as later waves of immigration from Asia and India), conducts a range of transactions, confrontations, brutalities, and occupations, to produce a complex cultural weave or chaotic composition.

Creole languages and voices negotiate an inherited violence that, within the context of the Caribbean, is specifically grounded in histories of the slave plantation. It is at this location, this site of stark abuse, that languages conflict and people violate others, to force a set of racial confrontations and cultural intersections. Glissant and other Caribbean thinkers, such as Kamau Brathwaite, have sought to emphasize how Creole languages, and *creolization,* arise from out of these confrontations, leading to frictions that are both "cruel and creative."[15] From within the deeply entangled complexity of colonial realities, creolization gives way to a lingual-politics – ways of speaking and writing *over* and *through* the dominant order. Such linguistic frictions form an arena that may aid in the creative expressions of a local voice, a lyrical work of the post-colonial subject.

Glissant argues for a "creolized" subject, and ultimately uses the term "a poetics of relation" to elaborate this. Here, poetics is specifically an "errant" operation, one that prolongs the intermixing and intersection of voices so as to capture their complex meanings; poetics is a relational construct, the formation of a language always already contaminated and bruised by multiplicity. Creolization is therefore understood not solely as a simple mixing, but rather one born from conflict and that results in the production of a poetic relation, an "inter-language" or oral mosaic of the migratory and the displaced, and which comes to embody "the chaos" of modernity.[16] The lingual-politics of such relations and the resulting narratives and musicalities – voices always already punctuated by the rhythms and mixes of creolization – instantiate a poetics whose multi-dimensional meanings unsettle the relations of master and slave, linguistic schooling and mother tongues, and the political ideologies of the cultured and the savage.

Creolization produces a voice of strangeness and estrangement, which forms the basis for trajectories of escape and empowerment; a *patois* whose expressions work against colonial languages by introducing an ambiguity of lexical origin, a broken tongue that, as Homi Bhabha suggests, may stagger the project of colonization through its uncertain mimicry: to appear as a copy, yet not quite.[17] To speak the words, but not quite properly. Through such poetic gestures, as Glissant further suggests, creolization is enabling the making of a global future, beyond that of the identity of the colonized (and the plantation system) and the lingering malaise produced by the foreclosure of a society to come. In short, the Creole is a subject of the future, one that announces through its performative figuring the global formations, transnational relations, and interlingual cultures of the contemporary; identities and languages brought into existence not only through speaking back, but *over* and *through* the project of modernity.

What Glissant attempts to elaborate, from within and out of this overarching history of violence, is the vitality expressed in the Creole. Creolization is posed as a practice of perseverance and resilience, a counter-linguistics that searches for a route past conditions of enslavement and loss. Through creolized poetics and vocalizations another journey is made, one that confronts the powers of the colonial project through an interweave of African homeland, European knowledges, the indigenous cultures and familial lineages, and the music and sounds of the islands, contributing to a set of powerful resources for nurturing an archipelagic orality.

> Pomme arac,
> otaheite apple,
> pomme cythère,
> pomme granate,
> moubain,
> z'anananas
> the pineapple's
> Aztec helmet,
> pomme,
> I have forgotten
> what pomme for
> the Irish potato,
> cerise,
> the cherry,
> z'aman
> sea-almonds
> by the crisp
> sea-bursts,
> au bord de la'ouvière.
> Come back to me,
> my language.
> Come back,
> cacao,

grigri,
solitaire,
ciseau
the scissor-bird
no nightingales
except, once,
in the indigo mountains
of Jamaica, blue depth,
deep as coffee,
flicker of pimento,
the shaft light
on a yellow ackee
the bark alone bare
jardins
en montagnes
en haut betassion
the wet leather reek
of the hill donkey.[18]

Derek Walcott, in the poem "Sainte Lucie," threads his way through a maze of shimmering words that criss-cross multiple languages and geographies. In doing so, he gives manifestation to the Creole sensibility or subject position I'm considering here, depicting and narrating the itinerant condition inherent to island culture. From the construction of a creole poetics, we may hear the staging of a form of agency, one that specifically skirts the logic of declaration and the master narrative. Rather, words come to occupy multiple positions, shifting between languages as well as geographies – a traversing of meanings and origins.

The question of new languages, and the possibility of speaking *over* and *through* histories of enslavement and colonial rule that Glissant addresses, is built upon the earlier work of Aimé Césaire, and the journal *Tropiques* which he co-edited along with Suzanne Césaire and René Ménil in the 1940s. *Tropiques* centered upon the general field of questions considered by Glissant, though from the perspective of the literary culture of the time (Martinique being under the rule of the Vichy government during the Second World War). Césaire captures this overall moment, notably in his seminal work *Notebook of a Return to My Native Land*, which remains a key text of Caribbean literature. The poetic prose of the work gives expression to the tensions at the heart of the colonial subject and is riddled with an embedded friction, a prose that veers back and forth between European and Creole languages, between the city of Paris (where Césaire studied) and the landscapes of Martinique, and between the traditions and folkloric imaginary of Africa as an original home and the contemporary scene of the island. Yet, within such a labyrinthine construct and reality we find the articulation of a complex questioning. Césaire grapples with his own existential position, searching through a surrealistic hall of mirrors for pathways from which to negotiate the question of one's own conflicted being.

Césaire partly answers these conflicts through the concept of *Négritude*. As a neologism, the term *Négritude* expresses the very journey Césaire attempts to make in his *Notebook*: a return that is never complete or truly possible, and which leads instead to the complicated project of new subjectivity, one that may, through a type of errancy, find the means to define oneself beyond colonialism. This new subjectivity is empowered through a confrontation with language – the French that is always already upon the tongue of the colonial subject. As Césaire argues, the subjectivity he seeks must work through the particular conditions of the islands and the making of new language.

> Whether I want to or not, as a poet I express myself in French, and clearly French literature has influenced me. But I want to emphasize very strongly that – while using as a point of departure the elements that French literature gave me – at the same time I have always strived to create a new language, one capable of communicating the African heritage. In other words, for me French was a tool I wanted to use in developing a new means of expression. I wanted to create an Antillean French, a black French, that, while still being French, had a black character.[19]

For Césaire, the project of a new language was guided by what he termed "poetic knowledge," which is enabling for opening a path toward a truly original and indigenous meaning. "Poetic knowledge," as he suggests, is based on the fundamental capacity of a word to bridge meanings in (racial) conflict, enabling a unique linguistic position by which to "conceive of the coexistence of opposites in the same term."[20]

The potentiality of poetic knowledge finds parallel in the oral mosaic expressed in Derek Walcott's poem above: a poetic knowledge born from errant subjectivity and that produces, like those diffractions recomposed on the islands, what Glissant additionally terms "écho-monde," or *echo-world*. From such an array of meanings, terms, and poetic imagery, we enter into an ecology of the senses and of the sensible; of errant subjects and oral mosaics, diversal positions and poetic knowledge affording the complex articulations of a black French, one that constructs a world in which words may carry opposing meanings simultaneously. Thus, Glissant's concept of the "echo-world," posed as yet another descriptive pathway into the maze that is the Caribbean, is an invitation to enter this deeply rhizomatic, heterogeneous archipelagic imaginary: a constantly shifting territorial arrangement from which new agentive formations are sought out – subjects born from echoes.

Formed by the conditions of displacement, these "echo-subjects" exist according to an ever-unstable origin, for we may never know for sure from where or from whom an echo first begins. Instead, the echo passes from one to the next, from language to language – from the *Irish potato* to the *pomme granate* – expanding as it goes and dizzying the certainty of any singular perspective. It is necessarily a transient agent, this echo-subject that fills the poem, passes from one to the next, is heard and repeated, sampled and mixed, to unsettle, through an inherent quality of itinerancy and passage – a "black fugitivity"[21] – the project of being a subject proper, a proper colonial subject.

The echo, in this case, is a subject of the world, set loose amidst language and culture, and that we might hear in lyrics by the calypso singer the Mighty Sparrow. For example, his poem "Dan is the Man" plays back, through a performative repetition, nursery rhymes taught to children in Trinidad as a way to "educate" locals on the English language and culture. From Brer Rabbit to Rumpelstiltskin, Gulliver to the Golden Egg, the Mighty Sparrow speaks *through* and *over* colonial linguistic schooling.[22] These are not forms of translation, or productions of a "linguistic hospitality" as Paul Ricoeur outlines as the basis for an ethics of mutuality.[23] Rather, these enunciations manifest inter-lingual frictions, performing a disobedient and fugitive lyric of negotiation and escape.

Creolization, practices of diversality and of echoing, suggests escape routes by way of inter-lingual manifestations and narratives. Accordingly, they produce movements not so much through acts of speaking back, but rather, by way of speaking *over* and *through*. Gestures of "critical mimicry," as suggested by Bhabha, are precisely the restaging of colonial cultures by local bodies, whose *replay* is never quite right, and therefore exceeds the limits of an appropriate rendition; a performativity, a *rhythming*, that works to slacken or tense the master language. In contrast to gestures of conscious appropriation or translation, such vocalizations extend and unsettle the culture of masters by moving *through* and *over* opposition and origin; *ingesting* foreign languages to bring them deep into the body, and to eke out a form of sustenance from the dominating culture in a complex move of incorporation and regurgitation, an echoing emerges that beats back the law with fugitive mixes.

Suzanne Césaire, in her essay "A Civilization's Discontent," summarizes the project of a Caribbean poetics when she writes:

> Let it be clearly understood. It is not at all a matter of a return to the past, of the resurrection of an African past that we have learned to understand and to respect. It is a matter, on the contrary, of the mobilization of all the combined vital forces on this land where race is the result of continual mingling. It is a matter of becoming conscious of the formidable mass of different energies that until now have been trapped within us. We must now use them in their plenitude, without deviation or falsification.[24]

These energies and continual minglings, giving way to echoing practices and errant rhythms, also produce mysterious and unlikely solidarities; by migrating across particular territories, trespassing and linking subjects scattered across parts of the globe, displacements and transience give way to new social formations - of shared speeches, common desires, causes driven by the meshwork of diasporic imagination and fragile alliances. This may extend, in the case of the *Négritude* movement, into aesthetic languages and shared sensibilities as found in the connections that were to form between Aimé Césaire and André Breton in the early 1940s. The poetical and critical writings produced by Césaire and his peers at this time, having found allegiance in readings of Surrealism, was to subsequently influence Breton when he discovered Césaire upon his arrival in Martinique while on his way to the United States in 1941. In this situation, the echoes that pass between poets are additionally shaped by the sharing of the French language; the colonial project, in scattering peoples and capturing subjects within an imperial grip, spills over and exceeds such systems to produce unlikely encounters, contaminations, and violations, as well as mystical and poetical correspondences. As Breton would write during his days in Fort Royal:

> Along bustling streets, beautiful faded polychrome shop signs exhaust all varieties of romantic lettering. For a moment, a sign held me under the same perverse spell as the paintings of René Magritte's negativist period. What do I ponder at a distance, is it an extremely ambiguous Magritte – is reality collapsing or creating itself anew? Just imagine a sky-blue butterfly, eagle-sized, on it the word PIGEON spelled out in white letters.[25]

Césaire and Breton's intersection, prefigured in words written and read from across the globe and yet in the same language, comes to manifest the deeply conflicted and complex reality of a certain modernity. These are echoes and poetics that *amplify* the colonial relation as one of "chaos" and "double consciousness"; the linguistic rhythms that beat back and forth across the Atlantic give way to errant (re)productions and the creative fugitivity of creolization.

Rasta Reasoning / "I and I" / Of Deep Echoes

The example of creolization, and the oral mosaic of diversality, locates questions of sonic agency within a specific context of colonial history. From within the mixed cultures of the islands, the violence of displacement has lent dramatically to emancipatory practices. By negotiating the power struggles over homeland, origin, and belonging, creolization and its literary cultures shift the terms of singularity toward those of multiplicity – a meandering and intersectional orality that sounds out other constructs of subjectivity.

The sound systems of Jamaica and the music of reggae, while shifting focus slightly, give cultural manifestation to the Caribbean diversality within which Creole languages and logics are to be additionally found. In this context, we find a number of performative modalities by which colonial relations are negotiated and brought into poetic and sonic structures whose rhythms and tonalities give way to visions of redemption as well as future homelands. As such, they deepen a view onto the itinerant and migrancy, to stage a dynamic sounded manifestation of how such conditions may be turned into forces of creative resistance.

Developed from the intermixing of European melodies and the rhythms of Africa, as well as the particular slave histories of Jamaica, reggae is fundamentally a music of protest and belief. It is a music concerned with "roots" and finds its power through a weave of social consciousness, a "versioning" of Christian belief based on the coming of the black Messiah, and the electronic possibilities that lead to the sound system, the sample, and the remix of songs and records. Here we find lyrics of prophecy and insurrection, a language of protest and festivity, and one that delivers a celebratory and dissident poetics based on the creative "reasoning" of the Rastafarian.

The lyrical and sonic force of reggae arises from out of the beliefs of the Rastafari movement, which emerged following the coronation of Haile Selassie I as King of Ethiopia in 1930. Rastas believe Selassie to be the second messiah, an incarnation of God ("Jah") whose reign signals an act of redemption from white domination. Drawing from Christian beliefs, Rastas enact a significant intrusion and "reasoning" onto the structures of European colonialism, shifting the narratives of salvation onto the islands of the Caribbean. Harking back to the earlier movement of Marcus Garvey, and his call for a journey back to Africa in the early part of the 20th century, according to Rastafari, Ethiopia operates as the national site for a physical and spiritual return.

In his book *Cut 'n' Mix*, Dick Hebdige highlights how reggae is full of dreams of black salvation. For example, "The story of Moses leading the 'suffering Israelites' out of slavery in Egypt is a particular favourite with reggae artists and audiences. It expresses the dream of black people in the New World - that one day they will be free enough to find the Promised Land again."[26] Rastafarian culture and reggae are marked by images of exodus and exile, by journeying and return. Displacement and dispossession, as well as migrations and itinerancy, are strong imaginaries within the lyrical constructions of reggae.

These expressions of "reasoning" are emblematic of the diversality expressed by Jean Bernabé et al. By repositioning the narratives of Christian salvation told through black identity, reggae punctuates Rastafarian culture with heavy bass and syncopated rhythms, and the prophetic power of sound to aid in one's journeys. Music and dance, songs and sound systems, are important cultural forces in Jamaica and across the Caribbean; they are carriers of stories and memories, spiritual narratives and prophetic callings. They express and give testimonial to island identities, and the archipelagic imaginary that is always, at the same time, bounded and woven into a greater constellation. Islands are sites of exile, of shipwreck and dispossession; people and things get stranded on islands, while at the same time they operate as important nodes within a larger network of trade and exchange, of passage and escape. In the historical context of the Caribbean, islands are bound to the directives of the empire, while remaining rather ungovernable; they are always potent sites of insurrection, pirate nations, mercantile and mercenary gangs and activities. In this regard, the archipelagic imaginary is one caught between exile and return, between enslavement and hybrid solidarities, inscription and the redemptive powers of song.

These are powerful images that lead to lyrics of resistance and hope, of prophecy and salvation, bringing us closer to a poetics of relation and the Creole sensibility Glissant describes. The deep voice of Prince Far I, for example, in narrating the Christian psalms in his album *Psalms for I* (1978), enunciates through a Jamaican *patois* a force of sounded subjectivity. His unmistakable vocalizations encapsulate the lost spirituality of the African home, told through the island culture of exile and return, to forge a "community of sentiment."[27]

The lyrical longing, based on spreading the word of the gospel for those islanders who cannot read, gives way to vocal practices and creole poetics, from "toasting" to "talking-over." Originating from the early sound systems in Kingston, where DJs such as Prince Buster would speak and shout over the sounds of records, the style of "talking-over" became a particular vocal practice. Talking-over and toasting are often based on the artist speaking and boasting of their superior qualities, of being King above the rest (a tradition that runs throughout genres of black music, from Little Richard to Eazy-E). Yet talking-over also captures the tendency toward sermonizing found in reggae. Prince Far I's deep talk-over resounds as testimonial pronouncements and prophetic callings, which gain traction through an indigenous musicality and sounded sensibility – a reasoning of not only black redemption and journeys to Africa, but equally through the echoic notion of "I and I," a spoken expression of Rastafarian sensibility replacing the single pronoun "I." The expression and idea of "I and I" indigenous to Rastafarian and reggae culture is based on the concept of oneness. Specifically, it refers to the oneness between two persons, and importantly, between oneself and God.

"I and I" as an expression is deeply suggestive for how we may hear within the *patois* and Creole of the Caribbean the articulation of a poetics of relation, a diversality by which I am always already more than one. This may be the spirit of God within us, or the solidarity that unites those in exile, or the self fragmented under the force of dispossession and colonization. "I and I" is a form of repetition, a subject position produced by a performative doubling, a *double tongue* making of oneself an *écho-monde*. "I and I" is a resonant proclamation by which one becomes many; a projection of solidarity and of power declaring that one is never alone.

The echoes and mixes of black French and Jamaican *patois*, the reasoning and versioning of Rastafarian belief, the rhythmic doubling and talking-over contorting the I as one of existential fragmentation as well as spiritual wholeness – these are modalities of powerful agency. The work of the poetics of relation constructs from loss and exile resistant imaginaries, routes along which new identities, alliances, and lyrical passageways are to be heard.[28]

Reggae additionally captures the force of this poetics by exploiting the transformative and transportable qualities of recorded electronic sound. Utilizing records as raw material for any number of mixes and remixes, reggae is fundamentally a practice of versioning, one that dislocates origin in favor of flexibility and transience, resilience and displacement, not to mention the disorienting hyper-production of delays and echoes central to the dub mix. Dub, in fact, is a form of mix saturated in delay and echo, giving way to hallucinatory works that deepen the vibratory drive and deep roots of reggae. From within the electronic delays saturating the music, one may detect the ecstatic arrival of a type of unification; according to a logic of displacement, of singularity always being prolonged into repetitions, of not quite the same – delays are not duplications, rather they spiral in and around origin, mutating as they go; from within this echo world of the "I and I" reggae culture constructs a form of wholeness, yet on its own terms: a communion with the spirit of Jah from which redemptive possibilities take flight. Echoes and delays are fundamentally acts of transmutation, giving support to the lyrical drive of island identity as it looks toward the horizon.

The practice of transposing the rhythms from one record to another, to be heard as an echo, ultimately modifies the notion of the original and recasts it within an altogether transformed sonic rendering - a freedom based on the repossession of one's own material culture. Language and the materiality of media are not so much fixed as immutable matter, as proper originality housed within a logic of Western colonial capitalism; rather, they give way through practices of diversality, of diffraction and echo, which radically express subject positions and sensibilities crafted from the poetic power found in the mix and the inherent migrancy of sound. Accordingly, the strictures and authority of ownership (which must be placed within a historical matrix of slavery and dispossession) are shifted, replaced by a creative "reasoning" and echoic practice that turn dispossession into a position of social and spiritual possibility. Within the dub of echoes and repetitions a listener may reclaim a form of cultural power and self-possession.

The mixology of reggae, the echoes and delays of dub, and the poetics of talk-over - the "I and I" that already displaces this body as an arrested subject: a type of archipelagic imaginary by which to skirt the grip of the colonial hand - these give manifestation to a sonic agency of the itinerant. Collaging assemblages of sounds, applying echoes and cuts, talking *over* and *through* the mix, such gestures function as powerful acts of a subjectivity whose echoes lead us back and forth across the Atlantic, from Jamaica to the UK, and further, toward a transnational and even biblical statehood articulated by the Rastafarian prophecy of a future homeland found in Zion.

I give an expanded description to practices found in reggae, and through Rastafarian and Creole subjectivities, in order to highlight from within conditions of displacement how forms of creative resistance may take shape. It is my view that, while embedded within histories of violence, the formations of post-colonial practices teach us much about crafting an art of survival. The testimonial and prophetic subject found in the enunciations of the Creole and the Rasta, and throughout other instances when voices figure ways to speak *over* and *through* histories of displacement, or from within the confines of adopted or imposed national and lingual states, force into view the complexity inherent to transnational relations. Itinerancy, accordingly, acts to condition the enunciations of a certain lingual subject, forming the lyrical vocabulary by which to construct a future to come – *to set the tongue moving*. As such, these are lyrics and vocalizations, rhythms and sounded acts deeply at odds with dominant structures and the law of the land, and yet they reverberate with great resilience, collecting through their fugitive performances a spectrum of poetic weapons.

The Migrations / And Desperate Walks / For No Borders

Might the lyrical actions of errant subjects give guidance for practices of the itinerant, and in support of those defined by migrations and displacements? For the establishment of certain vocal practices, subjects produced by *echoes* and *mixes*, and those that may incite ideas and narratives, a world-view of relations that enables a giving account of oneself as one journeys, and to which others may identify?

Sounds certainly move, but they also transgress, bundle, vibrate, filter, shatter, and penetrate; they may form into powerful cultural objects, recorded, sampled, cut 'n' mixed according to a project of errant subjects. Movement, therefore, occurs more as a constellation of motions and reverberations, all of which suggest forms of transient production as well as critical and creative trespass. Sound, in other words, is always *moving on*, or leaking out. Leaving so many sources behind, from bodies and objects to things and events, sound picks up and goes, and yet such going is not without its baggage or consequence. In *pushing on* sound collects a range of material elements – reflections, absorptions, reverberations ... possessions; these are *pressed* into the body of sound as it oscillates across and over surfaces. It is *bruised* by the environment, marked by the material features of surroundings around which it is shaped, impressed, sampled. By way of such gatherings and movements, sound may act as an extremely productive material and construct for nurturing acts of transgression and trespass.

As a sonic figure, the itinerant is a product of its surroundings and its travels; and what it carries forward is an assemblage of interactions. In this regard, it is a subject of the world, a foreigner with multiple languages, an *interlect* embodying the potentiality of a certain cosmopolitanism. Accordingly, an agentive position constructed from itinerancy may support connections and coalitions, *doublings*, across a range of locations, communities, voices, and histories. In this regard, it may greatly assist in the struggles embedded within the act of crossing borders.

Notions of creolization draw us closer to the agency of the itinerant by suggesting how to speak as, or to hear the voice of, not only the Creole subject, but equally that of the scattered and the lost, the body forced to move, to migrate and to flee, dislocated beyond the limits of the nation-state or homeland, or one haunted by past journeys and things left behind, as well as the possibility found in being out of place. Given the contemporary debates and conflicts surrounding the movements of refugees today, questions of colonial histories, bodies without place, languages of political representation, and struggles for citizenry all find new urgency. For example, the refugee movement in Berlin, which manifested in a dramatic stand-off with police in 2014 over the occupation of an abandoned school in Kreuzberg, centers upon the capacity to be heard, that is, to gain political space. Nadiye Ünsal, speaking as an active supporter of the movement, reflects upon this when she states: "The strength of the 'refugee' movement in Berlin still is that those threatened by the racist migration regime speak for themselves and confidently address the public."[29] Yet, speaking is not always easy, and certainly to find such confidence has required and continues to require other movements and tactics, gestures that force into public spaces the ongoing plight of migrants. This was given stark expression in the March for Freedom, which took place in the summer of 2014. Walking from Strasbourg to Brussels, asylum seekers and refugees were joined by supporters from across Europe to highlight their presence and struggle to the members of the European Parliament. Yet it was not only the destination, and the final demonstration coinciding with an EU summit on migration held in Brussels that mattered; as important were the stops and interludes along the way, where migrants and the undocumented camped to hold celebrations, to exchange information, to expose their legal struggles, their bodies and stories, as well as to nurture a new state of global consciousness with those encountered along the way.

Listening to the many different struggles means learning in its most practical form: from personal and collective experiences gained on the street, in protest camps, in direct challenges to an unjust border and asylum regime. Listening to and learning from those who refuse to accept the violent conditions imposed on them is inspiring and eye-opening. This process of listening and learning will be continued with the continuation, spread and intensification of struggles.[30]

Throughout the different protests and demonstrations across Europe, the refugee movement has sought to challenge the European "border regime" (Frontex) by fighting against deportation so as "to realize a community without borders."[31] Not necessarily a return to Africa (though that remains a profoundly deep narrative), but rather, to dedicate oneself to the future of a global culture by "disrespecting the borders imposed" on the undocumented.[32]

Migrancy draws forward questions as to the legal status of scattered people and what constitutes the right to movement. In the early part of the 20th century, vagrancy laws were equally debilitating to migrant workers who traveled across the United States, working harvests and infrastructure projects, among other labors. Traveling workers, or Hobos, were regularly arrested and subject to any number of abuses as the conditions of vagrancy cast them into a state of precariousness, leading to intense social and political conflict in the country (parallel to the labor movements that fought against the abusive system of factory work). This was additionally based on a political motive to literally house the Hobos, returning them to systems of wage labor which their itinerancy gave resistance to.

Essential to the "Hobo Army" (as the population of migrating workers were to become known) was a desire to redefine the conditions of labor, for instance by limiting the hours of a working week; and in addition, to challenge democracy and its relation to capitalism, a challenge which would find alliances across the world, in particular through the Industrial Workers of the World. This was given expression in a proposal by Irwin St. John Tucker, a socialist priest working in Chicago and active in the general circles of activists, sociologists, philanthropists, and clergy seeking to radically shift the conditions of workers and the poor in America. Tucker's proposal was focused on enlisting the Hobos into what he termed "The National Service Army,"[33] which would be in charge of "the national domain" – national forests and rivers, the common lands, and public resources. Tucker's proposal not only called for the new Army, but in doing so gave challenge to existing notions and stereotypes of the Hobo as lazy and worthless; instead, the

Hobo was situated as being central to the project of the nation, and in particular, a figure deeply aligned with public life.

To be without a home, as found in today's environment of not only refugees from abroad, but also homeless or evicted persons, forces one up against a framework of legality and rights, as well as the question of how we come to recognize and honor the needs of human life. The notion of home and of homeland is not only a powerful physical and psychic image, but also a defining feature by which citizenry and legal rights are gained.

How do we understand this voice then, this voice out of place and without legal status? How do itinerant voices resound within the contemporary Western environment? Explicitly those marked by African or Arabic origin, and yet grounded within European territories today? Meeting others, giving testimonial to their journey, and prophesizing about a future without borders - what are the lessons brought forward when encountering such vocalizations?

Vilém Flusser, in his critical meditations on what he terms "the freedom of the migrant," attempts to map out the potentialities inherent to being out of place (and further, to being without homeland, or "heimat"), and toward the conditions of exile and migrancy.[34] For Flusser, migrants radically disturb the "mysteries" embedded in the psychic imaginary central to the identifications one may have with national homeland; these mysteries spirit a deep sense for nationhood and national narratives by which one may feel located and integrated. In contrast, migrants, through their movement and displacement, which often include experiences of deep loss and fragmentation along with feelings of hope or ambition, must build their own narratives through creative and practical engagements with local conditions. This essential creativity, for Flusser, is a base from which belonging and place-making are made. Through interactions with an array of people, offices, and communities, and through finding or nurturing support structures with what can be found or gleaned, migrants force into being a new set of social and psychic formations. In this way, migrancy is a condition that undermines the "mysteries" by which national homeland is partly sustained. Instead, one is left to one's devices, and forced to reinvent meaning and relations from the often fragmented and challenging conditions of foreign territories. Migrants may certainly hang on to these mysteries Flusser describes, through a persistent identification and resistance to new local conditions, yet they may also spirit new constructs of community, thereby complicating and unsettling what counts as national identity and the maintenance of the properly local.

Flusser's thinking is suggestive for a theory of itinerancy, and for thinking migrancy and the potent disturbances it delivers, disturbances that may enable the establishment of new social formations, especially those grounded not only in narratives of homeland, but rather according to world-making activity. As he outlines:

> How can I overcome the prejudices of the bits and pieces of mysteries that reside within me, and how can I break through the prejudices that are anchored in the mysteries of others, so that together with them we may create something beautiful out of something that is ugly? In this sense each person who is without *heimat* has at least the potential of representing the awakened consciousness of all those who are settled in a *heimat*. He can be a vanguard of the future.[35]

Flusser's philosophical account of the migrant is one that aims to appreciate experiences of estrangement and alienation as not only negative or debilitating; rather, he embraces these as forces of social encounter, of new bonds and the crafting of resources, of knowledge shared and languages transmuted. Subsequently, such positions and productions may work to undermine or supplant the embedded violence of nationalism. Are not the foreign and the strange explicitly requiring another type of affiliation, exchange, and conversation? One that may spark a return to national limits (as seen in the arguments of the new right in Europe), but that may equally displace such limits? To incite new dialogue, often through the meeting of languages, around issues that also contain questions of rights and access? And that may renew gestures of social solidarity?

As with Glissant's poetics of relation, Flusser hears in the languages and voices of the migrant a productive inter-language, a voice whose itinerancy acts to extend us toward each other, and toward transnational realities. Returning to the articulation of the "I and I" of Rastafarian culture, I'm tempted to hear within it the break that Flusser speaks of – a break that is the migrant whose voice forces into the spaces of national culture a sound of displacement, a *rhythmed* orality: this voice that I may not understood fully, that speaks to me from a beyond suddenly so near, and that *percusses* the field of meaning. This broken tongue. The migrant voice is one that carries this doubling within it. The "double consciousness" of the diaspora, as Paul Gilroy outlines by looking at the colonial histories of the Atlantic in particular, is produced by negotiating between a past and a future, and a mixing of homelands, that ultimately affords a uniquely "transnational and intercultural perspective."[36]

The migrant voice as a doubling that produces its own echo, but equally a gap, a cut right in the center of subjectivity as it stands on this ground: in the pronouncement of "I and I" one speaks *toward* oneself, for one is always already elsewhere, formed from trajectories of displacement, and against which one must speak *over* and *through*. "I and I" is therefore a model of a political subject whose articulations we may hear as an attempt to speak beyond the violence of colonialism and toward a new state of self-determination as well as national identity.

We may hear such a production in reggae practices, but also in the gesture of the People's Microphone at play within particular occupations, notably during Occupy Wall Street in 2012. The process of one person's voice being repeated by a crowd so others may hear is a manifestation of a type of "echo-subject," a collective body constituted in the gaps generated by being expelled from the political. The People's Microphone amplifies not only the voice of a single speaker, but also that of an echo-world, one that may assist in overcoming the limits of the neoliberal city. Such echoes and collectivities resound as lyrical productions – the crafting of a collective subject – that may enable the crossing of those gaps which the dominant order works to entrench. In throwing the voice into the crowd, and into the bodies of others, the People's Microphone stages an inter-lingual voice, one that problematizes political speech as one of singular declaration, mobilizing instead the power of the collective subject and its echoic promise.[37]

This is not to belittle or overlook the power struggles inherent to collective resistances, or the limits that are continually arresting the opportunities for open assembly vital to democratic procedure or public power. Rather, I'm interested in casting the itinerant voice as one that reminds of the conflicted status of bodies without, which increasingly, as Glissant suggests, stands as the existential limit of contemporary relations of power. Diversality is therefore posed as a support structure, a coalitional base which may contribute, for instance, to the refugee movement's resistance to the "border regime" and lend amplification to its resounding lyric: "No

Border, No Nation, Stop Deportation!" For this lyric is heard as a composite of multiple voices, some accented and foreign – in fact, voices that attempt to constitute from their strained resonances a new narrative about citizenship beyond that of national borders are always foreign: this voice is never at home within closed constructs of singularity.

Returning to Glissant, it is clear that, while refugees from Syria and Iraq, for example, flee from the conditions of war, they deliver onto European territories a recognition as to the transnational, capitalistic, and imperialistic histories and realities that mostly determine such journeys. Accordingly, a poetics of relation may be haunted by conditions of war-torn devastation, torture, enslavement, and geopolitical economic projects. The violence of such realities though is not foreign to Glissant's proposal; rather, a poetics of relation is explicitly defined through an engagement with historical forces and the voices that struggle for recognition. For Glissant, the project of poetics is centered on overcoming what may divide, and in contrast, encourages principles by which to not only bridge cultural or political divisions, but also to understand these divisions as the basis for new languages, and ultimately, new global narratives. Subsequently, we may appreciate Berardi's call for poetry and Césaire's notion of poetic knowledge – resonant with Jean Genet's proposal that "poetic emotion" lies at the heart of revolutionary thought – as the basis for occupying thresholds and the liminal zones of dominant languages.

We may find this by returning to the Berlin refugee movement, which throughout 2014 occupied the Oranienplatz, in the neighborhood of Kreuzberg, forcing into view and into the media the presence of those without. In short, what the occupation achieved, to some degree, was to *politicize* their presence, reframing themselves as subjects with rights. "The Refugee Protest Camp at Oranienplatz has become a Germany-wide movement and is now connected all over Europe and beyond. So, the emergence of this self-confident and self-organized 'refugee' protest challenged the dominant, racist image about 'refugees' in Germany."[38] In her critical account, Nadiye Ünsal highlights the importance of this process of politicization, which contributed to shifting the discussion and activating protests toward formulating alliances across national and international borders – through occupations and speeches, but equally by walking over borders together.

Border Subjects / Giving Way to Border Publics / Encroachments

Migrations and evictions are deeply defining of the contemporary global environment in which mobility and expulsion are central features to a new logic, one of private debt, financialization, and competition, and technologies that enable all this on a new hyper-productive scale. Isabell Lorey gives further account of this new logic by highlighting the emergence of what she terms "governing through insecurity." According to Lorey, the neoliberal project is the realization of a biopolitical structuring under which "precarization becomes normalized."[39] From out of this framework, "Individuals are supposed to actively modulate themselves and arrange their lives on the basis of a repeatedly lowered minimum of safeguarding, thus making themselves governable."[40] The basic needs of living are no longer understood as fundamental rights, or placed within a structure of social security; rather, these are now the sole responsibility of private individuals and occur through conditions of ownership as well as debt. For Lorey, this new form of governing revolves around a logic of expulsion, where one is positioned upon a threshold to eviction.

A deeper consideration of migrations, of exile and displacement, is therefore extremely resonant with the stark realities of expulsion and insecurity that underpin the contemporary environment. Whether through diasporic poetics and the inter-languages of protest, shouts and the troubled syntaxes of the foreign voice as it speaks *over* and *through* new localities, or according to the occupation of abandoned buildings and acts of trespass, new emancipatory practices emerge to confront the question of borders and (in)security.

Kim Rygiel, drawing from research on the politics of migration, poses the notion of "bordering solidarities" to highlight those coalitions and communities that form around borders as they draw people in from either side; migrants and (non-)citizens, in particular, construct bordering solidarities in order to contest particular policies and the laws that infringe upon people's rights to movement and basic care.[41] For Rygiel, "borders also paradoxically can act as bridges or moments around which people on either sides [*sic*]" may come together.[42] Importantly, bordering solidarities unsettle the tensed inflexibility borders often perform, to incite debates onto the nature of their operations.

I would extend this notion, as Rygiel equally suggests, beyond that of national borders, and towards borders that run through any neighborhood or community, and that often appear within zones of conflict. Within recent struggles in the UK around social housing we find instances of bordering solidarity, for example in Newham, east London, where young mothers were evicted from their council housing due to budget cuts. In protest, the campaign known as Focus E15 seeks to make their cause public, undertaking particular actions that continue today in a struggle to secure new social housing. In a recent demonstration, which included the participation of an array of other protest organizations, the Focus E15 campaign occupied a former police station in East Ham and hung banners from its windows and ledges, one of which read: "No Room for Racism." As one protestor commented to the local press: "If they leave buildings empty like this police station – we will use them to make a political point. The housing crisis is driving people to despair and there should be no empty buildings whilst people are left to rot on our streets. We are also fed up of politicians blaming migrants for the housing shortage – it is just a dangerous lie – which is why one of our banners says – No Room for Racism."[43]

Forging links to other causes, crossing borders between citizens and non-citizen migrants, creating solidarities that locate the right to housing alongside the right to free movement, highlights how new formations of public power emerge to produce coalitional solidarities that fuel intensities of public discourse and acts of joining together. Mobility and migrations, housing and eviction, these form a complex thread running through the principle of governing through insecurity Lorey examines.

Precarity may be understood as the basis for not only new mechanisms of governability, but also what forces into action new articulations of citizenship and the formation of unlikely publics. As Glissant and Flusser suggest, the displaced exert a pressure onto the existing narratives and "mysteries" that define a nation and those granted rights from within or according to its borders. Instead, under the force of displaced subjects – of the evicted and the dispossessed, the exiled and the lost – new articulations of those borders are brought into play. From such instances arise languages and voices often never heard, and the exertion of powers that either support their reverberation or aim to subdue it. According to these struggles, borders take on new meaning, and other modalities of expressing political subjectivity and national identity are brought forward. In short, bordering solidarities work to give shelter against the prevailing logic of expulsion, creating new formations of community.

I want to carve out an additional perspective, to consider not only bordering subjectivities and their solidarities, but equally the drowned and the missing. These individuals and families that do not make it to the shores of Europe, and do not participate in the freedom walks or enter the squares of contemporary Western cities, but die attempting to do so. Are not such tragedies reminding of how the "space of appearance" is one deeply affected by those who never make it, by the lonely and the lost, the drowned and the disabled? In the voices and bodies that resound to one another, do we not hear, through a type of emptiness, the ones who never arrive? An emptiness that must additionally shape conversations, steering discourses toward the conditions that make such realities possible. In this sense, we must appreciate the forms of labor and struggle embedded in the conditions of public life overall, and the processes by which political actions are made, for it is not easy to arrive at situations where public discourses are generated, let alone heard. Counter to the notion of an educated public sphere, and persons freely entering into spaces of political action and informed by discourse, one may not even understand the language being spoken. Under such realities - which may refer us to any particular historical instance of mass migrations, or in relation to the uneducated, where we will likely find systems of abuse built upon the foreigner who does not understand, as well as those misinformed - what happens to concepts of the public sphere?

Asef Bayat provides an extremely useful view onto questions of dispossession and self-determination through his work on "nonmovements," which he defines as the "collective actions of noncollective actors."[44] In contrast to the organized and ideologically driven constructs of traditional social movements, Bayat turns to everyday practices, particularly within communities in the Middle East, by which ordinary people enact gestures of resistance, finding the means to sustain levels of freedom within systems of social and political control. These are gestures and practices that specifically elide the space of appearance (if any such space is actually available or imaginable), maneuvering through daily spaces and environments to eke out a livelihood, to form bonds with others, and to embody what Bayat calls "the quiet encroachment of the ordinary" – to intercede within daily structures, exchanges, languages, and rhythms.[45] Such encroachments form into a vague yet persistent cultural logic that defines the public sphere less through discourses and more through practices. These are "pragmatic politics" that shed ideological directives in order to construct ways of surviving, creating as well as taking pleasure in life lived with others. In so far as resistances are given articulation, these too are embedded within what can be made against the odds, joined together and constructed, make-shift or loosely formed, "overwhelmingly quiet" and that flow through daily exchanges.[46]

What constitutes the public sphere is thus never always clearly knowable or recognizable, nor does it remain static or without flexibility. Punctuated with civic generosity, precarious struggles, daily rituals between neighbors, multiple languages, economic disparities, and quiet actions – these are constituting practices and gestures that rework the public sphere in such a way that need not give way to overt political action or legible articulation. Rather, through pragmatic concerns and direct relations people may bypass what we understand as "political engagement" to nurture, instead, through acts defined by daily need and desire, life in the making.

* * *

Returning to issues of the itinerant, and the ways in which borders may create or require unlikely publics, we might recall that deeply complex instant, drawn from histories of the Caribbean, when during the Haitian revolt French soldiers were sent by Bonaparte to retake the island; as the invaders approached the island by ship, a faint sound could be heard – a recognizable melody, yet one that sounded slightly foreign, different than expected. This was the sound of Haitian slaves singing "La Marseillaise" in preparation for battle. We must linger over this instant, to hear its truly remarkable sounding, and its profound lessons. In such singing, we might hear the articulation of bodies defining themselves not only as "free subjects," but also as "citizens," yet a citizenship formed out of conditions of displacement and subjugation, inter-lingual and diasporic consciousness.

As the story continues, the French soldiers would hesitate, unsure of what to do: shall they drive on, to fight against people who return to them their own revolutionary spirit, through these resounding words of liberty? Or might they pause, to consider other possibilities, of solidarity or even retreat?

Borders are sites of contestation, enabling some while hindering others. They are lines that people struggle to cross and against which voices and stones are often thrown; they partially define identity while inspiring it to travel, to take flight – to imagine what may lurk beyond. They are powerful lines that inscribe themselves onto specific bodies and sensibilities, producing not only solidarities at times, but also "bordering subjects." The migrant, the exile, the displaced and the lost, along with those that reach across from the other side in support – these are bordering subjects, which may also act as bridges: people extending themselves, or stretched by national and geopolitical realities. These bridges enable the passage of not only solidarities, of knowledge and care, but also important lessons in privilege and loss, and what it means to be on either side of certain lines and languages. They are subjects with particular knowledge, archival bodies carrying certain histories and memories, and the intersection of cultural practices that have shaped their personhood, over and around borders. Bordering subjects are thus defined by itinerancy, by the echoes they produce, and as such, they embody the paradoxes and possibilities found in crossing borders or leaving home. As bordering subjects they are dangerous to systems that require of its residents, as AbdouMaliq Simone suggests, a resolute grounding. For the ones that dare cross over borders to enter where they do not belong inherently give threat to the maintenance of particular controls.

Accordingly, the voice of those who migrate and journey requires an extremely engaged type of listening, one that makes space for the displaced other. "So, we constantly have to control ourselves and also others to *not* talk *about* people *without* them. For many leftists this (self-)control is highly important. For others – due to lack of awareness regarding their own privileges and racisms – it is not."[47] Bordering solidarities and subjects are fragile bridges that require sensitivity for their own power relations, for these may unsettle even the "supporter" who tries to contribute by forcing oneself onto the space of appearance. For this space is never guaranteed nor does it hold steady according to a position of solidarity, or even a principle of moral right; rather, it is brought into play by those who claim it as an arena where rights must be argued for, and even desperately walked in support of – the essential rhythm of the body that is modulated, *remixed*, as a project of freedom. In this way, speech and action must be emphasized as inherently embodied, and often worked over as sonic and rhythmical matter. Whether particular subjects may perform a disquieting mimicry, or lift up through a surprising act of lingual identification the narrative of a nation, or struggle against the imposed frames of the stereotype or the scapegoat by remixing origins, speaking for oneself as a displaced and itinerant subject is often gaining momentum through critical rhythms that break the space of appearance and its borders.

Notes

1 See Saskia Sassen, *Expulsions: Brutality and Complexity in the Global Economy* (Cambridge, MA: The Belknap Press of Harvard University Press, 2014).
2 From an informal talk by AbdouMaliq Simone held at the ifa gallery, Berlin, 2016.
3 Sassen, *Expulsions*, 1.
4 See Franco Bifo Berardi, *The Uprising: On Poetry and Finance* (Los Angeles: Semiotext(e), 2012).
5 Ibid., 43.
6 See Stavros Stavrides, *Common Space: The City as Commons* (London: Zed Books, 2016).
7 Nick Couldry poses the need for new "intensities of listening" as a method for combatting the neoliberal logic of continual work. See Nick Couldry, *Why Voice Matters: Culture and Politics after Neoliberalism* (London: Sage Publications, 2010), 140.
8 Michael Warner, *Publics and Counterpublics* (New York: Zone Books, 2002), 114.
9 Ibid.
10 Berardi, *The Uprising*, 147.
11 Édouard Glissant, *The Poetics of Relation* (Ann Arbor: University of Michigan Press, 1997).
12 Robin Cohen and Olivia Sheringham, *Encountering Difference: Diasporic Traces, Creolizing Spaces* (Cambridge: Polity Press, 2016), 15.
13 See Jean Bernabé, Patrick Chamoiseau, and Raphaël Confiant, "In Praise of Creoleness." *Callaloo* 13(4) (Autumn 1990): 886–909.
14 Ibid.
15 Kamau Brathwaite, *The Development of Creole Society in Jamaica, 1770-1820* (Kingston and Miami: Ian Randle Publishers, 2005), 307.
16 See Glissant, *The Poetics of Relation*, and Bernabé, Chamoiseau, and Confiant, "In Praise of Creoleness," 6.
17 See Homi K. Bhabha, *The Location of Culture* (London: Routledge, 2004).
18 Derek Walcott, from "Sainte Lucie," in *Collected Poems, 1948–1984* (London: Faber and Faber, 1992), 310–11.

19 Aimé Césaire, quoted in Gregson Davis, "Forging a Caribbean Literary Style: 'Vulgar Experience' and the Languages of Césaire's *Cahier d'un retour au pays natal.*" *South Atlantic Quarterly* 115(3) (July 2016): 459.

20 Katerina Gonzalez Seligmann, "Poetic Productions of Cultural Combat in *Tropiques.*" *South Atlantic Quarterly* 115(3) (July 2016): 504.

21 The concept of black fugitivity is drawn from the work of Fred Moten, in particular from his book, *In the Break: The Aesthetics of the Black Radical Tradition* (Minneapolis: University of Minnesota Press, 2003). Here, Moten extends "blackness" (stemming from black musical practices of jazz and improvisation, for instance) as a powerful and empowering modality of being that works as an interruptive force, a "predisposition to break the law."

22 Sparrow (Slinger Francisco), "Dan is the Man," in Stewart Brown, Mervyn Morris, and Gordon Rohlehr (eds.), *Voiceprint: An Anthology of Oral and Related Poetry from the Caribbean* (Harlow: Longman, 1989), 130.

23 Paul Ricoeur suggests that translation performs a gesture of "linguistic hospitality" through its ability to mediate between different languages. For more, see Paul Ricoeur, *On Translation* (New York: Routledge, 2006).

24 Suzanne Césaire, "A Civilization's Discontent", in Michael Richardson (ed.), *Refusal of the Shadow: Surrealism and the Caribbean* (London: Verso, 1996), 99–100.

25 André Breton, *Martinique: Snake Charmer* (Austin: University of Texas Press, 2008), 55. As Breton would also write about Césaire's poetic project: "What makes [Césaire's demand for redress] priceless in my eyes is that it always transcends the anguish of blacks that is built into their fate in modern society and unites not only the anguish of all poets, all artists, and all the true thinkers, but through his verbal genius it embraces all that is intolerable and also all that is infinitely improvable in the human condition by our society."

26 Dick Hebdige, *Cut 'n' Mix: Culture, Identity and Caribbean Music* (London: Routledge, 1987), 49.

27 Arjun Appadurai, *Modernity at Large: Cultural Dimensions of Globalization* (Minneapolis: University of Minnesota Press, 1996), 8.

28 See Julian Henriques, *Sonic Bodies: Reggae Sound Systems, Performance Techniques, and Ways of Knowing* (New York: Continuum, 2011).

29 Nadiye Ünsal, "Challenging 'Refugees' and 'Supporters': Intersectional Power Structures in the Refugee Movement in Berlin." *movements. Journal für kritische Migrations- und Grenzregimeforschung* 1(2) (2015). http://movements-journal.org/issues/02.kaempfe/09.ünsal--refugees-supporters-oplatz-intersectionality.html (accessed March 2017).

30 The Struggles Collective, "Lessons from the Struggles: A Collage." *movements. Journal für kritische Migrations- und Grenzregimeforschung* 1(2) (2015). http://movements-journal.org/issues/01.grenzregime/19.from-the-struggles--lessons.html (accessed March 2017).

31 Ünsal, "Challenging 'Refugees' and 'Supporters.'"

32 From a poster produced for the March for Freedom. Downloaded from related website: https://freedomnotfrontex.noblogs.org/material/ (accessed February 2016).

33 See Irwin St. John Tucker, "Forward the Hobo! A Plea for a National Service Army" (undated), from the archives at the University of Chicago.

34 See Vilém Flusser, *The Freedom of the Migrant: Objections to Nationalism* (Urbana: University of Illinois Press, 2003).

35 Ibid., 15.

36 See Paul Gilroy, *The Black Atlantic: Modernity and Double-Consciousness* (Cambridge, MA: Harvard University Press, 1995), 15.

37 See Jeremy Woodruff, "A Musical Analysis of the People's Microphone: Voices and Echoes in Protest and Sound Art, and *Occupation I* for String Quartet." PhD thesis, University of Pittsburgh, Department of Music, 2014.

38 Ibid., 5.

39 Isabell Lorey, *State of Insecurity: Government of the Precarious* (London: Verso, 2015), 11.

40 Ibid., 70.

41 Kim Rygiel, in Ilker Ataç, Anna Köster-Eiserfunke, and Helge Schwiertz, "Governing through Citizenship and Citizenship from Below: An Interview with Kim Rygiel." *movements. Journal für kritische Migrations- und Grenzregimeforschung* 1(2) (2015). http://movements-journal.org/issues/02.kaempfe/02.rygiel,ataç,köster-eiserfunke,schwiertz--governing-citizenship-from-below.html (accessed March 2017).

42 Ibid.

43 See https://focuse15.org/2016/07/12/police-station-targeted-by-housing-activists/ (accessed July 2016).

44 Asef Bayat, *Life as Politics: How Ordinary People Change the Middle East* (Amsterdam: Amsterdam University Press), 2010.

45 Ibid., 14.

46 Ibid., 19.

47 Ünsal, "Challenging 'Refugees' and 'Supporters.'"

5

The Weak

The gestures that bring us into the world, making us seen and heard, felt and witnessed by others, carry with them an intrinsic forcefulness; we gesture, we move, we impinge onto our surroundings and others. These are embodied actions by which subjectivity gains definition and which produce effects and meanings from their intensities – they make impressions from which certain consequences are generated. These actions, and their repercussions, are formed by and through the interactions and reactions between oneself and others – I am only myself in so far as others shape me, and through which I in turn shape others. The deep and defining relationality of being a subject in the world, however, is in constant tension with given social institutions, with the processes of language and the limits of speech, and through one's access to support structures, including the extremely important and highly varied relationships of which one is a part. We are not only relating body to body, subject to subject, but equally according to the institutional frameworks that enable or limit contact, movement, and responsiveness. I *feel* myself with and through others, as well as by entering or exiting the institutions and offices of society – by scraping across the limits and structures of the social order and the permissible.

Relationships, in this way, mostly extend from the personal to the institutional, creating a more entangled and experiential way of being in the world in which moments of exchange, sharing, and feeling are shaped by particular frameworks. In turn, one contributes to those frameworks, demanding entry and participating in their activities, bending the languages and the practices that perpetuate particular institutional orders. From such perspectives, the sensual nuances of feeling and of being felt, of wanting and needing, greatly inflect the actions and gestures that make one seen and heard in public life, and that inform how such visibility and audibility lead to or hinder positions of social and political participation.

Audre Lorde, in an essay on "the erotic," suggests that it is by way of the sharing of joy that the productive conditions for mutuality and empowerment may be nurtured. As she describes: "The sharing of joy, whether physical, emotional, psychic, or intellectual, forms a bridge between the sharers which can be the basis for understanding much of what is not shared between them, and lessens the threat of their difference."[1] Lorde furthers this thinking by arguing for a richer integration of joy and pleasure – our excitements and our vitalities – within the institutional constructs of family relations, work, and public life. The sharing of joy thus acts as a highly charged foundation from which forms of cohabitation, interpersonal exchange, and mutuality may emerge. Importantly, Lorde poses the erotic as that which acts to bridge the "spiritual" and the "political." From such a condition, she writes, "we begin to feel deeply all aspects of our lives, we begin to demand from ourselves and from our life-pursuits that they feel in accordance with that joy which we know ourselves to be capable of."[2]

The erotic, the sensual, and the joyful come to act as an empowering basis for putting into practice the intensities central to being a thinking and creative body, specifically by supporting our inherent desire to "share" in such intensities: the erotic is first and foremost a generative project, born from touching and being touched, by the depths of a sensuality that also, importantly, forces us into a state of vulnerability and interdependency. Lorde's "erotic subjectivity" is one that situates the affective and sensual state of personhood in and around the political, and is thus the beginning of confronting all that oppresses or dominates – that thwarts the full blossoming of life's vitality – whether in the form of institutional systems or through the internalized fears and anxieties that keep one locked within limited conceptions of oneself and others. The joyfulness of erotic becoming, as the bristling of life with others, ultimately leads to a "visibility which makes us most vulnerable" and yet which is also "the source of our greatest strength."[3]

The experiences of joyful sharing, of the fullness of agency and actions – the activeness of the breathing and feeling body that is ourselves and that flows from us and along with others – such a position supplants theories and systems that would separate politics and the personal and, by extension, the public and the private. In contrast, for Lorde, and for others, in particular bell hooks, whose ideas continually seek to bridge life lived and the formations of public representation (political and other), spaces of political visibility are greatly influenced by the psychological, emotional, and relational passions by which one experiences and desires from the world and others. From such a position, is not the political a space of relations and mutuality served not solely through reasonable deliberation or strategic alliance, but one equally shaped and instituted by what *moves* this body? By the intimate relationships and emotional knowledges that often sustain communities, and that become central in times of conflict?

bell hooks captures the question of "passionate politics" by arguing for a mode of coming together "in that site of desire and longing" which may act as "a potential place of community-building."[4] hooks is dedicated to steering questions of politics according to the affective lessons of desire and longing, as well as through an ethos of loving relations. Through her work on education and practices of pedagogy, hooks furthers her line of thinking about social and political struggle, and the importance of nurturing a more holistic approach to critical thinking and community-building; reflecting upon her own experience as an educator, she writes: "As a classroom community, our capacity to generate excitement is deeply affected by our interest in one another, in hearing one another's voice, in recognizing one another's presence."[5] Seeking to nurture a learning community, the power relations between teacher and student within the dynamic of the classroom must give way to more socio-communal exchanges, which are nurtured by caring for everyone's participation. The "engaged pedagogy" hooks works toward is therefore dependent upon continually recreating the conditions of the classroom according to the life experiences, the desires and fears, the ambitions and the fluctuating moods of teacher and student alike.

Following Paulo Freire's statement that "education is an act of love,"[6] hooks' demand for a more rigorous integration of one's emotional and personal knowledges into the classroom is paralleled in Lorde's concern for the erotic. Lorde works to put into question the "false dichotomy" of the spiritual and the political, which results "from an incomplete attention to our erotic knowledge." "Our erotic knowledge empowers us" to take responsibility for the world around us, to deepen, through "the passions of love," understanding and engagement with the intensities of life with others.[7] The erotic, as that vibrant and tensed bridge between the spiritual and the political, echoes with the "commonality of feeling"[8] hooks attempts to create in the classroom. Shared spaces of erotic becoming, sites of desire and longing, bridges between private and public life, these concepts argue for modes of subjectivity that are equally discursive and emotional, reasoned and felt, driven by individual passion and the collective intelligence of communities. Political determination and struggle is thus housed alongside the greater joys of erotic sharing, which must be understood to flush public discourse with intimate contact and the frictions of comingling.

* * *

Movements and propagations, oscillations and trespasses, sounds may deliver powerful energies to annoy and to interfere, to agitate and to violate, yet such powers are fundamentally based upon a condition of diffusion and dissipation. Sound is always moving *away* from a source; it abandons origin, it longs and is perennially leaving. In traveling and migrating, in brushing up against numerous surfaces, being absorbed or reflected as it moves, it is equally losing weight, shedding identity. It is thinned out as it goes. As it migrates, invading any number of territories, to sweep past and through the social field, brushing the skin and contouring the rhythms of places – sound does so according to a condition of weakness. It is, as a defining feature, a weak object – *how can I hold it, this sound?* Additionally, it spreads such weakness. To listen one must pause, even stop what one's doing; we fix our ears to certain sounds, or we even block them in moments of invasive noise. We are touched or hit by sound as it brushes passed or burrows deep within, to send us to sleep, or to fluster and flush the cheeks; it makes us move, toward a point of exhaustion and exhilaration. We are pushed around by sound, and accordingly, we often weaken, losing our energy and tolerance, or our ability to sit still. How impatient we are when asked to listen. How vulnerable we are to the force of a sonic event, to the comings and goings of sounds. We are both uplifted and annoyed by sound, by the tonalities and the vibrations, the songs and their repetitions.

One is nurtured by the humming of certain melodies, sheltered by an ambient whirl of oscillations, all of which exist on an unsteady threshold that may give way to sleepless nights or distressed conversations.

The unsteady and highly charged experiences of sound and listening refer us to what Didier Anzieu calls "the sonorous envelope," which enfolds us at an early age to charge our first perceptions with degrees of intensity. It is from the sonorous wrappings, which purr around us, that we draw out our first experiences of being a body in the world, and importantly, how the stirrings of sound work as a relational medium or channel linking the deep rustlings of the interior to the motions and movements of the external world.

> Before the gaze and smile of the mother who feeds and cares for it, reflecting back to the child an image of itself that it can perceive visually and internalize in order to reinforce its Self and begin to develop its Ego, the bath of melody (the mother's voice, the songs she sings, the music she lets it hear) offers it a first mirror of sound, which it exploits first by cries – which the mother's voice reacts to with soothing noises – then by gurgles, and finally by playing with phonemic articulation.[9]

Sounds, following Anzieu, dramatically contribute to the emerging features of subjectivity, producing bodies and selves sensitive to the primary flows and forcefulness of sonority as a link to the animations around us. Importantly, such experiences lead to a deep sense for both the soothing and abruptness of sound, and the ways in which it may weaken us. The sonorous envelope, by bathing and trembling us with its oscillations, leaves a deep impression upon the psyche, placing sound within a matrix of sensuality, desire, and psychic intensity.

The weakening experiences that sound may wield can be highlighted by considering the phenomenon known as Autonomous Sensory Meridian Response (ASMR). ASMR is founded on inciting euphoric sensations by way of auditory stimulation or triggers, mostly based on quiet rustlings, soft whisperings, and textured scrapings. This has led to an active online community, particularly through the circulation of YouTube videos produced by ASMR practitioners. Often employing a binaural microphone and headphones, which allows for an enhanced stereo image, these videos aim to induce a heightened state of euphoric listening, one whose experience often leads to meditative states, "tingly" feelings, and sleep.

ASMR, while referring us to a particular cultural community, may introduce the sonic agency of the weak that I'm pursuing here. Weakness, as I'm keen to suggest, captures a positive potential, one that stands in contrast to normative representations and narratives of the powerful and the strong, the virile or the stiff. As such, to grow weak at the knees, to faint or to be unable, may act as an alternative framework for modalities of being a subject in the world, not to mention emergent forms of resistance and mutuality.

Weakness is not only articulated through abusive forces that may fix one within a system of dominance; rather, it is equally an essential human condition, articulated in moments of crisis, fragility, and loss as well as through "joyful sharing" and the simple instances of feeling oneself touched by another; a vulnerability central to being human. These conditions and experiences may additionally act as the basis for countering systems of domination. Following Lorde and hooks, weakness may deliver a performative impasse in which powers of dominance stagger, and from which forms of self-determination and shared resistances may find traction. To let oneself go within the fervor of joyful contact, or to grow limp in the arms of another, a friend or even a police officer when refusing to vacate – *this body carried off* – or to resist through silently standing still, or holding firm together, these expressions find their strength not only through political conviction, but through a deep appeal to moral conscience.

Although weakness is undoubtedly tied to experiences of torture and abuse, or cast pejoratively toward the lazy and the disabled, I'm interested to consider how weakness may form the basis for undermining systems of abuse by appealing specifically to an ethics of interdependency and the non-violent. And further, to what Václav Havel calls "the power of the powerless" – weakness as a resistant stance, an articulation of responsibility and conscientiousness, and one that may instantiate another understanding of strength, what I may call "weak-strength."

What types of emancipatory expressions might weak-strength contribute to? If resistance and insurrection require formations of assembly, acts of disobedience, and continual pressures exerted against systems of subjugation or injustice, how might weakness operate as a means or tactic? How may the silences of refusal and passive resistance embolden social movements and community-building? And more broadly, how might we articulate or expose our weaknesses as affirmations of life?

To pursue this questioning, I want to consider a number of historical examples, as well as related philosophies and protest communities, in particular those based on non-violence, conscientious objection, and a loving ethos. This will provide a base for exposing the degrees to which speaking and listening often entail confrontation and negotiation with regimes of violence and abuse, especially as a means to nurture processes of open dialogue. As Susan Bickford suggests, practices of listening are not necessarily based on friendship, or even affability; rather, within processes of deliberation and discourse, listening works to enrich dialogical argument and disagreement.

In such dialogical situations, how might strength operate? What resources can we draw upon to take up a listening position, for sustaining open dialogue in the face of abusive power, or even vindictive speech? In what ways might joyful sharing and the affective politics of the erotic appear in situations of struggle? Strength, in this instance, may be found in the ability to foster the conditions of dialogue, to give recognition through listening to others passionately. Strength is therefore expressed by holding the tongue, waiting, and giving space to the voices of others – through the ability to foster a collective acoustics of interlocution and mutuality. Such an acoustics though must be underscored as the basis for a counter-speech – to say No to others. In this regard, the nurturing of the conditions of dialogue must also include the possibility of articulating one's refusal to listen.

Ecstasy / Collective Vibrations / By Which to Weaken / And Raise

Weak-strength is posed as an ethical base from which to honor the voice of another, and the dialogical necessity central to what it means to live, work, and feel together. Yet weak-strength, as found in the dialogical instant or relational arena, may equally be considered by attending to alternative practices and scenes; not only as an operation within the time and space of open dialogue, but one that figures prominently within situations of conflict and contestation, and through embodied expressions of shared ecstasy and passionate politics.

A prominent example that may assist in detailing practices of weak-strength can be found, as part of a much greater counter-culture movement, at the anti-war demonstration in 1967 held in Washington, D.C. Organized by the National Mobilization Committee to End the War in Vietnam, along with activists Jerry Rubin and Abbie Hoffman, the anti-war demonstration assembled together counter-cultural communities alongside the old and new left, and can be seen as the concretization of an emergent culture of political agitation. Shifting emphasis from traditions of worker and union strikes central to the old left, the demonstration instead found recourse through alternative references and actions, in particular, expressed in the idea of levitating the Pentagon. The idea was to gather together the energies of a range of metaphysical resources and spiritual traditions, and to channel these, through an act of collective chanting, against the military compound in order to "exorcise the demons."[10] This included a series of ritual gestures performed around the Pentagon, for instance acts of purification, surrounding the Pentagon with cornmeal, as well as reading aloud what Ed Sanders, a member of the band The Fugs, called an "exorgasm" text.[11] The counter-culture movement thus gave expression to a new imaginary of public power through creative gestures of free love, conscientiousness, and cosmic vibration - what Abbie Hoffman would term "the politics of ecstasy."[12]

Theodore Roszak, in his study on the counter-culture, points out how the movement understood clearly that protest and resistance needed to be directed beyond any single figure or office of power. Rather, the entire social and political structure had to be challenged, striking "beyond ideology to the level of consciousness" so as "to transform our deepest sense of the self, the other, the environment."[13] Through a politics of ecstasy and consciousness, the movement set out according to a far-reaching, imaginative, celebratory and moral vision of transformation, to which the idea of levitating the Pentagon gives expression.

In parallel to Hoffman's call for a politics of ecstasy, the question of raising consciousness is a defining feature of the women's liberation movement, and in particular, was expressed in the formation of "consciousness-raising" groups at this time. Initiated, for example, by the New York Radical Women in the late 1960s, the studying and questioning of women's experiences, and discussions of literatures on pertinent themes such as childhood, employment, and motherhood, formed the basis for the movement. Kathie Sarachild, a founding member, captures the ethos of the project when she writes: "Consciousness-raising – studying the whole gamut of women's lives, starting with the full reality of one's own – would also be a way of keeping the movement radical by preventing it from getting sidetracked into single issue reforms and single issue organizing. It would be a way of carrying theory about women further than it had ever been carried before, as the groundwork for achieving a radical solution for women as yet attained nowhere."[14]

Radical politics was thus grounded in a greater project in which people's lives, and the institutional structures around which they are situated, were central and pressing; discourses and actions, assemblies and coalitions, personal behavior and thinking together, these were all interconnected, leading to what Sarachild succinctly called "the radical weapon" of consciousness-raising.

The politics of ecstasy and of consciousness-raising form the basis, albeit for a diversity of reasons, by which to confront the technocratic and patriarchal conditions of the system at this time; a response to the currents of social and political life that, following the aftermath of the Second World War, instated a systemization of production and consumption centered on the emerging new order of global capital. The United States, in taking center stage within this order, came to enact an increasingly imperialistic engagement globally (which the Vietnam War exemplified). Such growing dominance – bolstered through economic, industrial, and military development – was to lend dramatically to a new society of affluence as well as alienation.

Herbert Marcuse's studies at this time, which found a receptive readership in the counter-culture movement, are extremely poignant and suggestive for understanding the degrees to which alienation expressed the underside to the growth and prosperity of the nation and the West in general. The first line of *One-Dimensional Man* (Marcuse's popular analysis of industrial society) already announces the critical position the author would stake out: "A comfortable, smooth, reasonable, democratic unfreedom prevails in advanced industrial civilization, a token of technical progress."[15] Marcuse's "one-dimensional man" is posited as the paradoxical result of Western affluence, where thought and speech are constrained by a "technical organization" of society in which production and consumption do less to nurture social relations, and a deepening of shared resources, than to shield us within a logic of progress as well as paranoia. Against the prevailing "technological reasoning" of the times, which reduces the creativity of speech, for example, to an overarching "functionality," restricting the full "dimensionality" of discourse, Marcuse aims to support the "emergence of a new historical subject."[16] Subsequently, the author provides both insight and vocabulary for countering "unfreedom," and its related "administration," which would equally be taken up by the counter-culture movement and challenged through questions of ecstasy and consciousness.

Constrained by "one-dimensional thought" and an overarching focus on technological progress, the counter-culture sought an extremely different way of life, one driven by a radical individualism guided by moral concerns for peace and love. As the collective act of levitation suggests, the counter-culture was a movement in how bodies feel, perceive, and act together. This would give way to understanding how political subjectivity is expressed not solely in gestures of speaking up, or in rational collective assembly, but equally in "arational" formations of energetic attunement, ecstatic togetherness, and affective intervention - vibrational formations in which the personal is deeply political and the political is something to be lived and shared. The counter-culture thus provides a compelling expression of an insurrectionary sensibility that works to unsettle the state of political functions, and the dominance of technocracy, infusing speech and action with wild imagination and a sonority of ecstatic being.

The counter-culture's project of struggle was shaped by an appeal to the senses and ideas of radical sharing arrived at by way of cosmic as well as communal resources. Vibrations and reverberations, the consolidation of ecstatic energies and conscious thoughts, these gave support to a prevailing desire to not only intervene within particular policies, but to counter the ideological directives embedded within the technocratic order. In short, upon the psychic labors by which bodies live, dream, relate, and join together. Psychedelic music, flower power, free love, group therapies, women's liberation, and communalization, for instance, all sought to reorganize consciousness and bodily freedoms, sexuality, race relations, and conventions of feeling and thinking, which would impact upon institutional structures – that could shake the structural foundations of society. New formations of living, often gravitating around co-operation, as well as self-organized and communal structures, were centered within the project of revolution.

The focus on enacting social transformations of living led to a range of strategies and expressions, and found partial guidance through mind-altering practices, notions of dropping out or tuning in, free speech and thought, and the creative instituting of alternative lifestyles. In short, gestures and actions that may support radical consciousness, to give formation to the "bridges" between the spiritual and the political articulated by Audre Lorde. Although there are clear differences between struggles for freedoms as they appear in black communities, white suburbs, in colleges or in rural areas across the United States at this time, Lorde's "erotic subjectivity" is deeply suggestive for describing the ethical and moral revolution initiated overall by such struggles. On a fundamental level, the focus on new states of self-determination and community realization brought forward a revolution in how particular bodies would come to confront one another; the erotic dimension that Lorde seeks to capture is one that inflects these confrontations with the deep promise of joyful contact, which may include the tensions inherent to "passionate politics," but equally the power of mutual recognition.

The state of erotic subjectivity finds manifestation in aspects of the counter-culture movement and its dedication to new states of consciousness and bodily freedoms. The relation to hallucinogens, for instance, while problematic, is suggestive for the new modalities of consciousness that were to enable a politics of ecstasy. The general disorientation central to hallucinogenic experiences is one that disperses identity, diffusing one's sense of self into an associative framework in which delineations between interior and exterior, bodies and things, inorganic and organic dissipate. Through disorienting practices and affective attunement, might the spiritual and the political find new points of contact? In the throes of disoriented being, one-dimensional thought may explode into a thousand rays of joyful sharing: to *weaken* us as a project of deep becoming.

Aldous Huxley's influential study on altered states of consciousness, *The Doors of Perception*, first published in 1954, is suggestive for the possibility of new consciousness. Based on a series of mescalin experiences, Huxley sets out to notate the subsequent effects. What follows is a meditation on perception and consciousness, which leads to a recognition of all that interrupts the profoundly limited concept of the "I." In contrast, under the influence of mescalin, perception opens out to what Huxley terms the "living light" of the "Not-self" from which the world is no longer ordered according to "distances and measurements," but rather brims over with flows of intensity. Objects, materials, and things in the world take on a profound singularity of being in which "The mind does its perceiving in terms of intensity of existence, profundity of significance, relationships within a pattern ... of living light."[17] Sitting in his home in Los Angeles, Huxley absorbs the world around him and gives description to a primary state of erotic subjectivity: "The legs, for example of that chair – how miraculous their tubularity, how supernatural their polished smoothness! I spent several minutes – or was it several centuries? – not merely gazing at those bamboo legs, but actually *being* them – or rather being myself in them; or, to be still more accurate ... being my Not-self in the Not-self which was the chair."[18]

The intensity of confronting directly the vitality of material presence – that things are not so much inert matter to which one attributes meaning as a resource, but rather are housed within a mutuality in which things *and* oneself are intertwined upon a plane of relationality, cohabiting the "living light" from which meaning arises – provides an explicit education on how to renew worldly contact and togetherness. The world, in short, appears as a common experience in which all things matter and from which one-dimensionality erupts into multiplicity.

Timothy Leary's quasi-spiritual acid sessions, conducted ten years after Huxley's experiments, would galvanize the hippie movement around the notion of transcending one's individual ego – to experience "ego-death" in order to be reborn into the "clear light of a thousand rays," as Leary would proclaim. "There are no longer things and persons but only the direct flow of particles."[19] The potential of rays and ecstasies are clearly what Rubin and Hoffman, and all those that chanted outside the Pentagon, sought to capture and influence: to shift the formation of particles that make up the military-industrial complex of the nation and to weaken its foundations through a vibrational assembly, not only a collective voice speaking forth, but a collective, sonic reverberation guided by the living light of new perceptions and the shared consciousness of the Not-self.

Acts of weak-strength, and expressions of "ego-death," appear throughout the anti-war resistance, supporting counter-cultural thought in which hallucinogens, meditation, spiritual communing, political organizing, musical culture, street graphics and fashion, tuning in and dropping out, all rotated around a moral and philosophical agenda aimed at shifting the cultural ethos from the technocratic to the organic, from imperial bombast to conscientious objection, from missiles to flowers. Ego-death was to parallel the spiraling numbers of dead US soldiers (and Vietcong) reported weekly on various television channels, not to mention the surge of dead musicians, community leaders, activists, and politicians. Death, loss, ecstasy, rage, and love were to produce a cultural era of great conflict and transformation, hope and violence. Within such a context, hippies struggled to hang on, and the non-violent passions of counter-cultural subjects and civil rights activists became beacons of hope, as well as articulations of ongoing ideological conflict.

The Pentagon demonstration in 1967 produced not only a confrontation with US military policy, but equally the emergence of a new sensibility and the articulation of how that sensibility may enact social transformation. As Abbie Hoffman would ecstatically proclaim:

> We will dye the Potomac red, burn the cherry trees, panhandle embassies, attack with water pistols, marbles, bubble gum wrappers, bazookas, girls will run naked and piss on the Pentagon walls, sorcerers, swamis, witches, voodoo, warlocks, medicine men, and speed freaks will hurl their magic at the faded brown walls. Rock bands will bomb out with "Joshua fit the Battle of Jericho." We will dance and sing and chant the mighty OM. We will fuck on the grass and beat ourselves against doors. Everyone will scream "Vote for Me." We shall raise the flag of nothingness over the Pentagon and a mighty cheer of liberation will echo through the land.[20]

The powers of this collective will expressed outside the Pentagon is additionally depicted in the iconic photograph by Bernie Boston. Titled *Flower Child*, Boston's photograph captures a young man as he inserts a single carnation into the rifle barrel of a National Guardsman outside the Pentagon. The relation between the protestor and the guardsman gives expression to notions of weak-strength that I'm interested to consider. In this instance, forms and manifestations of weak-strength work to not only resist regimes of violence, but base such resistance upon a responsibility for life in general. Exposed weaknesses deliver onto the scene of political struggles a moral imperative, one deeply aligned with the heart.

Parallel to the Pentagon demonstration, the civil rights movement equally set out to challenge the establishment of the United States' political system by arguing on the basis of constitutional rights. Yet those rights, for instance the right to vote, were articulated within a larger reference to inalienable rights and an appeal to human dignity and moral responsibility. Although the anti-war movement was not necessarily a movement on the question of rights, it did significantly align itself with a greater wave of resistance based on equality, and a concern for human life. Boston's photograph captures and becomes emblematic of a greater moral sensibility and debate, and partly gains its sense of righteousness through the vulnerability expressed on the part of the protestor, this unarmed figure that stands with a flower as his only weapon. How might a flower stand up to a rifle? Or further, what particular political realities make it necessary to pose such a confrontation? What forces can be channeled and brought to bear against the military complex of the nation, here exemplified by a rifle at the ready and one pointed at its own citizens?

Boston's *Flower Child* is emblematic of the anti-war movement, equally capturing the emergence of Flower Power, which took off in the summer of 1967 in San Francisco. The use of flowers as forms of resistant weaponry was initially suggested by Allen Ginsberg in his essay "Demonstration as Spectacle as Example, as Communication," from 1965. Here, Ginsberg urges protestors to confront and surpass the "war psychology" by handing out flowers.[21] Subsequently, flowers circulated, were brandished and offered during demonstrations and marches, along streets and throughout daily life, gestures which found their full blossoming in the summer of love. The image and use of flowers became indexes for the growing anti-war and counter-culture movements, and eventually for the articulation of Flower Power as an ethos of loving relations. If the Vietnam War forced into view a highly public manifestation of military power, shaped through a national paranoia around "the red scare," Flower Power countered such a view with an altogether different political practice, that of conscientious objection, pacifism, the non-violent, and an ethos of peace and love embodied in the hippie whose masculinity, for example, contrasted sharply with that of the traditional soldier.[22]

In what way might social transformation be directed through expressions of ecstatic being and holistic communing? Might weakness lead us to the limits of a certain political arena, exposed not so much by bodies in movement, but by those standing still, or by those chanting collectively? Consciousness-raising and passionate politics refer us to questions of emancipatory struggle by grounding debate within a framework of moral conscience – a concern for the world and for others. We might equally glimpse such concerns within more recent social movements and protests, from the Occupy and Gezi Park protests to commoning practices and informal pedagogical projects, which stage not only important forms of resistance, but equally express a contemporary imagination for embodying new political subjectivity. These are expressions that not only stand against certain policies, but do so through an appeal to new broader social formations in which the political is always already a question of feeling and of being felt.

As with the emergence of Flower Power in the late 1960s, contemporary struggles are not solely issue-based, but rather derive their intensity from an overall appeal to moral objectives, erotic subjectivity, as the capacity to share differences, and earthly responsibility, and that question the forms of representation and institutional structures that mostly govern community passions. The "commonality of feeling" bell hooks articulates, as that which supports deeper relations that specifically bridge intimacy, family relations, and friendships with that of institutional and political life, is expressed not only in the sharing of words and discourses; in contrast, a commonality of feeling is reliant upon emotional and personal knowledges, spiritual guidance, and the capacity to empathize.

hooks' project, which is often relating itself to questions of education and of raising consciousness, finds resonance in contemporary struggles that call for a deep shift in political structures and social solidarity. As has been noted, Occupy's "leaderless" quality, along with its plurality of demands, for example, points toward a general and pervasive indignation for the conditions and structural capacities by which politics works on behalf of people today.[23] Instead, a "communal and democratic reorganization of public life" has emerged from out of the occupation of "world squares."[24] Directed towards civic engagement, the occupations are equally about processes by which individuals may feel themselves as part of what matters. These are protests and initiatives aimed not solely at renewing political life, but further, in support of the desires and longings for joyful sharing.

Non-violence / Standing Still / Within Churches from Below

The life and work of Martin Luther King, Jr., shaped by a dedication to non-violence, was to exert enormous influence throughout the struggles of the 1960s. Grounded in Christian values of love and charity, as well as the lyrical and redemptive rhetoric of the black church, King became an essential driving force in the struggle for civil rights. The relation between political protest and moral conscience articulated a powerful position, a stance taken as an intervention within the space of appearance, one that specifically brought black individuals and communities into the open (and often at great risk). The founding principles of the United States, based on liberty and equal rights, were re-articulated from within the realities of discrimination, forcing into the public sphere the historical violence of slavery and its present-day institutional expressions. Segregation and racial discrimination became, accordingly, moral issues that brought into the open a deep questioning as to the ways in which founding principles and constitutional rights were to be lived and upheld. Within this sphere of conflict set against the backdrop of the 1960s, the ethos of non-violence, along with expressions of passive resistance, wielded a strong alternative vision for equal living.

Principled beliefs, moral imperatives, and passive resistances were cast as a means to transcend the social fabric of segregation and prejudice, and further, to challenge a political system articulated through technocracy as well as war and violence. It was by way of consciousness and conscience that new social and institutional formations were sought out, new solidarities nurtured, and even political changes inspired. In this case, weak-strength, in the form of passive resistance and non-violent civil disobedience – along with the politics of ecstasy and consciousness-raising – operated as a powerful platform, enabling a reinvention of political subjectivity and practices of resistance; to sit in, to hold together, to love openly, to speak freely, to march forward in song, to construct sites of "desire and longing," and to be vulnerable, such formations and modalities of assembly found strength precisely by refusing to give in to campaigns of violence.

Non-violence and expressions of weak-strength were to appear within cultures of resistance worldwide, and often suggested transnational links that might equally turn into a global movement of shared moral passions and beliefs. The case of East Germany (and by extension Eastern Europe) is helpful to consider, as it is based not only upon individual commitments to egalitarian living, but the ways in which particular institutional coalitions may spearhead non-violent resistances.[25]

In the setting of East Germany throughout the 1970s and 80s, formations of resistance against the dominating Socialist Unity Party of Germany (SED) found support from church pastors who sought to establish more independent and direct dialogue with local communities and the issues that concerned them. For instance, environmental issues became increasingly pressing in the context of East Germany, as manufacturing industries and the general state of food production were leading to an intensification of pollution and ill health. In this context, in the late 1970s a group of young activists attempted to mount protests and demonstrations focusing on the environment, events which were subsequently put down by local police. In response, activists found refuge in the Zion Church in East Berlin where, in collaboration with church leaders, they set up a library dedicated to environmental issues. Known as the Environmental Library, it soon became a meeting point for peace and environmental activists, who would exchange information and printed materials gathered from around the Eastern bloc, organize the printing of the underground journal *Umweltblätter*, as well as discuss the general conditions of socialism in East Germany and engage in a range of actions to contend with the state system. In this context, churches throughout East Germany played an important role in being one of the only institutional frameworks in which civic and political concerns were given outlet and guidance. Importantly, the setting of the church, along with its commitment to giving shelter to the needy, lent significantly to the overall project of resistance; practices of free thought, environmental responsibility, and civic generosity were thus housed within a larger framework related to moral conscience and non-violence, as well as the idea of a "Church for Others" that had become prevalent in Germany.[26]

Throughout the 1980s increasingly defiant pastors also independently organized themselves, and eventually declared a "Solidarity Church" as a means to construct a more autonomous position from that of the state church, which was under constant pressure from the East German government to participate and generally fall in line with "socialist" principles. The Solidarity Church, in contrast, aimed to challenge notions of a "church under socialism" with a "new grassroots movement for a different socialism."[27] In the summer of 1987 this led to organizing a "counter-synod" to the official one being convened by the Evangelical Church. The counter-synod, which went under the name "Church Congress from Below" (*Kirchentag von unten*), was based on remodeling ideas of socialism in "accordance with certain humanist ideals."[28] Such a declaration also shifted the discussion from questions internal to the church and toward those of human rights and abuses of power in society.

The Church from Below, as it became known, fought for independent activity and thought, forming a rare culture of spiritual and political resistance contoured by the church as an environment of care and compassion. The church movement ultimately sought to create new conditions for socialist living deeply resistant to the hierarchical, dominating structures of the East German government (which included the brutalities of the Stasi). Subsequently, this led to intensified efforts, and in the fall of 1987 the Zion Church organized the first unofficial peace march. This led to numerous raids by the East German police on churches throughout the country, and included the temporary closure of the Environmental Library and subsequent arrests of peace activists. As a result, tensions increased and actions of resistance were further mounted, including the intervention within a parade to celebrate Rosa Luxemburg and Karl Liebknecht, founders of the German Communist Party, where protestors unfurled a banner with a quote from Luxemburg which read: "Freedom is always only the freedom to think differently."

In addition to the Zion Church, the Gethsemane Church in East Berlin, and St. Nicholas in Leipzig, in particular, became focal points for the peace movement and important sites in which the eventual stand-off between East German police and protestors would dramatically contribute to the opening of the wall in early November 1989. Encouraged by the ongoing shifts under Gorbachev in the Soviet Union, and his policies of *Perestroika* and *Glasnost*, churches and the oppositional groups they came to house organized peace prayers, hunger strikes, candle-lit marches and silent vigils in support of social reform, as well as the release of demonstrators from a prison in Leipzig, corresponding with the fortieth anniversary celebrations of the GDR. In particular, the Gethsemane Church became an important base for the intensification of resistance throughout East Germany. Guided by an ethos of non-violence, people found sanctuary and support in the space of the church, and in particular, through what became known as the Monday prayer meetings. Oppositional groups and activists gathered there, utilizing the church telephone with its long-distance capability, which enabled the collecting of information from around the country, and occupying nearby streets. The East German police reacted violently, beating and wounding demonstrators, and arresting hundreds. This led to an overall increase in protests, all of which were marked by an unwavering commitment to non-violence on the part of demonstrators. "No Violence!" could be heard resounding through the streets, a refrain that came to extend the ethos nurtured from within the church out into the city.

Extraordinarily, over the course of intensified demonstrations the undercover agents from the Stasi, who regularly infiltrated the Monday prayer meetings throughout the year and the related vigils, would change sides, lending their support and sympathy to the resistance movement. Christian Führer, pastor at St. Nicolas in Leipzig, commented: "That was an incredible event. We could never have reached so many party members through any written effort or by any other means all at once, and something in writing would not have accomplished this; one simply had to have been there and experienced it."[29]

The experience of the Monday prayer meetings, and a dedication to freedom of thought and non-violent resistance, were crucial to the culture of the peace movement; gatherings shaped by whispered reports, non-violent ethos, shared books and discussions, hunger strikes, peace workshops, and the moral belief in human rights – a righteousness whose vocalizations would find amplification not only in speaking up and out, but in the silence of the vigil and the sitting still central to refusal, acts that, following the resignation of the leader of the SED, Erich Honecker, in mid-October 1989, would greatly contribute to the wall being opened a few weeks later.

Václav Havel, in his important essay "The Power of the Powerless," written in 1978, examines the conditions of Soviet occupation, or what he terms the "post-totalitarian system." According to Havel, the Soviet system represented a new configuration of power and governance, in which the system itself functions over and beyond any single individual or leader.

> Part of the essence of the post-totalitarian system is that it draws everyone into its sphere of power, not so they may realize themselves as human beings, but so they may surrender their human identity in favour of the identity of the system, that is, so they may become agents of the system's general automatism and servants of its self-determined goals.[30]

Post-totalitarian systems ultimately subject one to a larger ideological apparatus which, according to Havel, essentially envelopes one's existential and spiritual being, directing all of one's imagination and desires toward a single directive: the perpetuation of the system itself. In this context, one is subject to the force of a "lie," which constrains the rich and dynamic tendencies of a life lived openly and "in truth." As Havel suggests, "While life ever strives to create new and 'improbable' structures, the post-totalitarian system contrives to force life into its most probable states."[31]

Within the conditions of post-totalitarianism, life goes elsewhere, into "hidden spheres" of activity and collective work, partially enabling the unfolding of life lived in truth. Václav Benda, a contemporary of Havel, would articulate the idea of a hidden sphere through an additional concept, that of a "parallel polis." As Benda wrote in 1977: "I suggest that we join forces in creating, slowly but surely, parallel structures that are capable, to a limited degree at least, of supplementing the generally beneficial and necessary functions that are missing in the existing structures, and where possible, to use those existing structures, to humanize them."[32] The parallel structures, for Benda, would function to counter the post-totalitarian apparatus, not only by enabling certain activities, but importantly, for nurturing a broader moral resistance.

Havel's understanding of the hidden sphere ultimately leads him to the idea of a "post-democratic" system, one arising from out of the collapse of Soviet control (as well as any post-totalitarian system) and shaped by what he already perceived occurring in the parallel polis of the Czech underground: an "existential revolution."

In other words, are not these informal, non-bureaucratic, dynamic and open communities that comprise the "parallel polis" a kind of rudimentary prefiguration, a symbolic model of those more meaningful "post-democratic" political structures that might become the foundation of a better society.[33]

The hidden sphere, a parallel polis, counter-cultures, and churches from below, these perform according to an imperative of resistance that gains traction through a greater moral gravitation and concern. As Havel suggests, it is only through the conditions that enable one to live life in truth that we may nurture the founding of a better society. Such desires and concerns, while aligning with particular Catholic movements in the Czech resistance, equally found recourse through expressions of counter-cultural communities, in particular, the musical festivals and theatrical happenings, for example, that were to galvanize the Czech underground (and the writing of Charter 77[34]). Within these contexts, these scenes of dissident gathering and creative expression, one was always acting as an exposed subject, a body open to arrest and abuse. Strength was thus to be found by holding firm to a larger moral and cultural orientation, the nurturing of a "solidarity of the shaken,"[35] and the making of a "commonality of feeling."

The contexts and situations that I'm considering here, which are produced by communities under duress, oppressed and therefore driven to forge cracks and cavities in the state system in order to join together, may refer us back to bell hooks and her notion of the sites of desire and longing; hidden spheres and churches from below, while defined by political resistances, are in turn sustained and nurtured from the joyful sharing of collective vitality: the music festivals and social gatherings in the countryside outside of Prague were essential celebrations, where music and theatricality enriched the dissident discourses and resistant communities which were to prove important in the transition to democracy in 1989. Additionally, both the Zion Church and the Gethsemane Church, as the driving centers of resistance in East Berlin, are located in the neighborhood of Prenzlauer Berg, which was also the center of the artistic and cultural scene of the city. As Carlo Jordan, a co-founding activist of the Environmental Library, suggested, it was partly due to the cultural vitality of the neighborhood that these sites became forces of political change.[36] Powerlessness is therefore often gaining strength through the production and sharing of cultural and creative expressions, which are first and foremost the articulation of an erotic force and which work to stage forms of rehearsal for a society to come.

Listening / Loving Relations / This Rage / And Other Weapons

In her work on democratic culture and citizen practices, Susan Bickford posits that listening is fundamental to supporting and experiencing conditions of dialogue and deliberation; that listening performs to secure processes of assembly and decision-making central to democratic societies. As she suggests, listening requires patience, and the stillness of a body, in order to foster processes of exchange, in which exposure and even vulnerability are supportive of dialogue. Yet listening is not without its challenges. Speech makes a claim onto an existing situation and the attention of others. In speaking, I take up space, and in doing so I may perform certain privileges of which I may not be fully aware. In speaking, I may reinforce an existing imbalance or discriminatory condition, for "what tends to get heard in public settings is a way of speaking associated with those who control social, political, and economic institutions."[37] One is vulnerable to the voice that speaks forth, and yet one must also give room for what another says and the force of their narrative or argument, for "we have the capacity to hear something about the world differently through the sounding of another's perspective."[38]

In this way, listening requires a condition of weak-strength; within the dialogical moment I am always listening beyond myself, moving my own views in consideration of another's, giving my attention to opinions different from my own while finding ways to resist and counter their power if need be. Listening is never purely passive, rather it performs as an affective and intelligent labor by which recognition is nurtured and relations are continually remodeled. For Bickford, the importance of listening is found in its capacity to potentially "break up linguistic conventions and create a public realm where a plurality of voices, faces, and languages can be heard and seen and spoken."[39]

The plurality of voices that may create a highly dynamic public realm though seems to suggest more than rational debate and reasoned deliberation; while I may sit patiently as another speaks, I am equally pressed and strained by what I hear; I fidget in my seat, I'm agitated by the slowness of the process – I may, in fact, hate what I hear. And in speaking, I search for a logic that may equally move others, to craft a narrative from which agency may be nurtured, even inspired to go beyond the current state of affairs. Voice and listening, in other words, are also shaped by passions and imagination, by desires and longings, by rage and frustration, which fill our words and our listening with intensity. It is precisely this intensity that Nick Couldry searches for when he calls for a renewal of "spaces of politics" within neoliberal society, and which I detect in counter-cultural subjectivity and resistances from below.[40] Yet, I would also supplement Couldry's call by not only emphasizing "the political," but equally what Havel terms "the pre-political."[41] For Havel, the pre-political may guide us according to the essential needs and desires of the human condition, to redirect constructs of power through moral urgency and responsibility.

Havel's shift toward the pre-political as the basis for post-democracy echoes with what bell hooks calls, following Martin Luther King, Jr., the "beloved community" – that what I hear and say are fundamentally shaped by the force of what it means to be a human subject with others. Is this not what Abbie Hoffman imagined when he stood outside the Pentagon making jokes and trying to craft the ecstatic joys needed to start the revolution? And what Audre Lorde suggests by way of bridging the spiritual and the political through the guiding knowledge of the erotic? A community founded on principles of love?

Judith Butler, in her recent work on precarity, offers a compelling path for expanding processes of speech and listening, and which may aid in specifically confronting regimes of violence. In response to practices of torture by the United States military following 9/11 and the subsequent Iraq War, Butler attempts to understand how it is possible that, on one hand, the US may espouse a moral responsibility to uphold and defend democracy on a global level while, on the other, continually perpetrate crimes against humanity. As has been blatantly revealed with the circulation of the Abu Ghraib photographs (of US military abusing prisoners), along with subsequent reports, practices of torture and human rights abuses run throughout recent US military actions.[42] In light of this, Butler is concerned to reinstate a deep sense of moral responsibility onto the debates surrounding ongoing US interventions and their victims. In doing so, she raises a number of compelling questions, for instance: what makes one human life more grievable than others? How might the ways in which certain bodies are deemed "less than human" be challenged, to remind of a greater shared humanity and responsibility?

Accordingly, she poses the condition of "precarity" as that which pervades human life, underscoring vulnerability, interdependency, and mutual care as essential factors for sustaining life in general.

> The apprehension of the precarity of others – their exposure to violence, their socially induced transience and dispensability – is, by implication, an apprehension of the precarity of any and all living beings, implying a principle of equal vulnerability that governs all living beings.[43]

If one is able to recognize the inherent precarity of one's own life, and if one may then recognize the degree to which one's life is dependent upon others, might this shift the conditions that make possible extreme practices of abuse, intolerance, violence, and war? As Butler suggests, it is within practices of torture that the "body's vulnerability to subjection is exploited" and "the fact of interdependency is abused."[44] Yet, precarity and vulnerability are also conditions for responsiveness, for the "formulation of affect" and from which new recognitions and responsibilities may emerge.

> Our obligations are precisely to the conditions that make life possible, not to "life itself, " or rather, our obligations emerge from the insight that there can be no sustained life without those sustaining conditions, and that those conditions are both our political responsibility and the matter of our most vexed ethical decisions.[45]

Following Butler's deeply engaged work, and Bickford's reflections on listening, precarity functions as the basis for shared responsibility and moral compassion, which not so much lessens the possibility of standing up or attacking systems of violence, but rather supports bodies in movement and those that aim for resistance by way of loving relations. In other words, our weaknesses may be mobilized as potent weapons – what James C. Scott succinctly terms "weapons of the weak."[46] One such weapon of weakness may be found in the time and space of listening, and nurtured through the diffusiveness and potency of silent actions, collective vibrations, and the co-soundings of shared passions.

I'm reminded of the action of Erdem Gündüz, the performance artist who, during the protests in Istanbul in 2013 over the Gezi Park incident, took to the streets to stand silently. Throughout the protests, Gündüz appeared, standing motionless; this mysterious figure seemed to attend to the powers at work around him through quiet contemplation, a slow and concentrated focus. His steady, silent figure came to act as a form of protest, persistently confronting the clashes around him by contrasting the eruption of anger and conflict with deep silence. Gündüz's demonstration soon became a galvanizing force, and was replicated by others who would stand beside him for hours. Such actions, forming a weapon of the weak, and reminding of the ways in which weakness – here expressed through a steady silence, an unwillingness to act out, or to speak up, and yet moved by the desire to be present, exposed – may deliver a deeply moving and affective expression, giving suggestion to what Asef Bayat terms a "revolution without movement."[47]

From a series of investigations and reflections on practices of social resistance, Bayat considers the ways in which "revolutionary projects" at times are shaped and initiated not through actions of "frontal attack," but rather through daily practices, gestures of co-operative assistance between neighbors, and the consolidation of resources within a community. These "nonmovements" are not so far from current situations and experiences of political conflict; the range of "revolutionary projects" or insurrectionary communities gaining traction from the fall-out of austerity measures and global crises equally point us toward what Bayat terms the "encroachment of the ordinary": formations and initiatives that bypass party politics to undertake forms of direct action, shared responsibility, and daily practices of co-operation. Here, encroachments of the ordinary intervene within the social and political landscape to produce commonalities of feeling and strategies of working together.

Returning to issues of weak-strength, and the project of raising consciousness as a radical weapon, I want to consider how encroachments of the ordinary take shape through an overall ethos of loving relations. The social transformations generated by the emergence of digital technologies and their incorporation into the private sphere over the last 25 years have clearly altered relations between the private and public spheres. With the incorporation of practices of work and the "productive" activities of public positions into the domestic environment, the traditional distinctions of private and public have been reoriented to produce new forms of livelihood, family structures, and professional identities – we are less and less distinguishing between work and personal life, for example. From these new configurations, the experience of the private sphere of home-life, as being tied to acts of caring and nurturing, of reproduction, is no longer clearly separated from the productive actions of public work and exchange. Accordingly, the interweaving of reproduction, and of building home, with that of worldly production, of labor and business, effectively shifts the nature of political behavior.

The shift in private and public relations, in short, ignites new understandings in what it means to act politically, and what we may expect from those that come to represent our personal concerns. Instead, a new spirit of political subjectivity and civic imagination is demonstrated throughout recent protest cultures, such as the Gezi Park protest and the Occupy movement, as well as by the nonmovements of everyday life. Within these examples are to be found an intensity of political behavior that radically addresses the overall sphere of life: questions of economic policies are equally questions of domestic life, and actions against governments are shaped through projects of self-organized care and neighborhood initiatives aimed at nurturing from the ground up the life of others.

The spirit of political behavior today gives articulation to a concern for common resources, open media platforms, shared responsibilities and earthly care, and work to resist the "frontal attacks" in favor of coalitional frameworks composed of a diversity of views and issues, people and places. From such positions and expressions I'm interested to consider in what way acts of care-taking, family work, and home-building may come to inform politics today, one that extends the reproductive labors of the home as the foundation for an ethos of loving relations and generosity *onto* the world. Such a shift is already deeply informing the emergence of protest cultures today; as seen in the Gezi Park demonstrations, the issue of cutting down trees in order to erect a shopping mall mobilized a mass protest and ultimately gave way to new expressions of civic mindfulness and environmental responsibility: the subsequent "movement of the parks" has fed into "communal activities such as the planting of collective gardens and shelters and the caring of street animals."[48] The shift toward conscientious objection and projects of radical care is extended from greater historical narratives of resistances and protest cultures in which questions of civil rights, gender equality, environmental sustainability, and moral responsibility led to attempts at usurping power structures often defined by imperialistic domination.

What types of "public spheres" emerge from these transformations of political behavior, which demand or inspire an ethos of loving relations? Might the new languages attached to the public imagination today, by interweaving the reproductions of home life with the productions of worldliness, support sensibilities of care and intensities of compassion? And which, by extension, may nurture community-building, cooperation, and publicness that is equally about home, family, care-taking, and well-being?

Although the political is generally understood not to be the time and space for nurturing intimacy and erotic subjectivity, it may in fact be what is needed in today's environment; as the dynamics of governing power reach into the essential conditions and experiences of what it means to act within the world, shaping bodies and lives, livelihoods and future hopes, through neoliberal and illiberal agendas, practices of political life that engage an ethics beyond the purely political seem necessary. It is precisely this question that Havel answered by way of what he calls the "pre-political," which he understands as the fundamental moral responsibility of life lived with others that override party politics. This view is echoed poignantly in notions of intimacy and love driving the work of bell hooks, which argues for an understanding of public life that includes not only the exposed conditions of vulnerability and interdependency, but also how such conditions are often expressed through intensities of rage. For to be weak is also what may drive us toward resistance – to desperately seek alternatives to one's own precarity, and to demand one's right to human life in its fullest. Rage, in fact, is a potent source for driving forward the desire to transform the conditions of life surrounding and from which to fight for what one loves most.

Weakness is an occasion or experience of being overwhelmed; it reveals us in instances of collapse and deflation, a condition of loss or surrender that equally reveals our interdependencies – our essential bonds. It reveals us at our most vulnerable, a body without and in need. In relating to the fundamental condition of weakness, I've sought to question how political behavior may lead to a more sensible relation with others, a sensibility of deep recognition and moral compassion equally grounded in the passions of living life with others. In bridging the spiritual and the political, might new expressions of protest and resistance turn us toward arenas of public discourse and community-building from which politics and an ethos of loving relations may work side by side? Weakness – as that shuddering of erotic subjectivity, a precarity that requests of the other a deeper listening, a more considered touch – is cast as a powerful axis around which "sites of longing and desire" are formed; in being weak, one is in need of others. Accordingly, as Judith Butler suggests, the constructs of power may be grounded to give way to greater consideration for the responsibilities of oneself. As examples of resistance and protest attest, speech and action may gain moral direction by tuning to the ecstatic presence of another's touch and the social circles dedicated to deepening consciousness.

Notes

1 Audre Lorde, "Uses of the Erotic: The Erotic as Power," in *Sister Outsider: Essays and Speeches* (Berkeley: Crossing Press, 2007), 56.
2 Ibid., 57.
3 Ibid., 42.
4 bell hooks, *Outlaw Culture: Resisting Representations* (New York: Routledge, 1994), 217.
5 bell hooks, *Teaching to Transgress: Education as the Practice of Freedom* (New York: Routledge, 1994), 8.
6 See Paulo Freire, *Education for Critical Consciousness* (London: Bloomsbury Academic, 2013), 34.
7 Lorde, "Uses of the Erotic: The Erotic as Power," 56–57.
8 hooks, *Teaching to Transgress*, 8.
9 Didier Anzieu, *The Skin-Ego* (London: Karnac Books, 2016), 186.
10 See Norman Mailer, *The Armies of the Night: History as a Novel, the Novel as History* (New York: Penguin, 1968).
11 The text reads: "In the name of the amulets of touching, seeing, groping, hearing and loving, we call upon the powers of the cosmos to protect our ceremonies in the name of Zeus, in the name of Anubis, god of the dead, in the name of all those killed because they do not comprehend, in the name of the lives of the soldiers in Vietnam who were killed because of a bad karma, in the name of sea-born Aphrodite, in the name of Magna Mater, in the name of Dionysus, Zagreus, Jesus, Yahweh, the unnamable, the quintessent finality of the Zoroastrian fire, in the name of Hermes, in the name of the Beak of Sok, in the name of scarab, in the name, in the name, in the name of the Tyrone Power Pound Cake Society in the Sky, in the name of Rah, Osiris, Horus, Nepta, Isis, in the name of the flowing living universe, in the name of the mouth of the river, we call upon the spirit to raise the Pentagon from its destiny and preserve it." http://wagingnonviolence.org/feature/the-day-they-levitated-the-pentagon/ (accessed January 2017).
12 See Abbie Hoffman (Free), *Revolution for the Hell of It* (New York: Dial Press, 1968).

13 Theodore Roszak, *The Making of a Counter Culture: Reflections on the Technocratic Society and Its Youthful Opposition* (Garden City, NY: Anchor Books, 1969), 49.

14 Kathie Sarachild, "Consciousness-Raising: A Radical Weapon," from a talk given at the First National Conference of Stewardesses for Women's Rights in New York City, March 12, 1973. https://organizingforwomensliberation.wordpress.com/2012/09/25/consciousness-raising-a-radical-weapon/ (accessed October 2016).

15 Herbert Marcuse, *One-Dimensional Man: Studies in the Ideology of Advanced Industrial Society* (Boston: Beacon Press, 1964), 1.

16 Ibid., 252.

17 Aldous Huxley, *The Doors of Perception* (London: Vintage, 2004), 9.

18 Ibid., 10.

19 Timothy Leary, with Ralph Metzner and Richard Alpert, *The Psychedelic Experience* (London: Penguin Books, 2008), 106.

20 Hoffman, *Revolution for the Hell of It*, 39–40.

21 See Allen Ginsberg, "Demonstration as Spectacle as Example, as Communication, or How to Make a March/Spectacle." http://voices.revealdigital.com/ (accessed February 2017).

22 Flower Power contributed to a cultural attitude of tolerance and acceptance, finding recourse through an engagement with and appropriation of Zen Buddhism. Ginsberg's example is built upon a legacy, from the Beats onward, in which Zen and Taoist thought offered an alternative and guiding framework. Ginsberg's flowers, Kerouac's dharma bums, John Cage's acceptance of all sounds, Allen Watts' holistic philosophy, Eric Fromm's new political psychology, Agnes Martin's minimalist paintings, or Mira Schendel's rice-paper drawings, these are clear instances from which we may understand a larger cultural project throughout this time, one whose ambitions were based on escaping or undoing ideologies of the post-war technocratic society. Conscientious objection and passive resistance were thus politicized expressions of this culture, whose projects were increasingly founded upon a belief in moral being and interdependent relations.

23 See Mikkel Bolt Rasmussen, *Crisis to Insurrection: Notes on the Ongoing Collapse* (Wivenhoe: Minor Compositions, 2015), 79.

24 Mehmet Döşemeci, "Don't Move, Occupy! Social Movement vs. Social Arrest." *Krytyka: Global Activism*, Special Issue (2014): 16.

25 It is interesting to note that Martin Luther King, Jr. visited West Berlin in 1964 to commemorate the life of President Kennedy, who had recently been killed. King's presence in the divided city galvanized the East Berlin church communities, and through informal channels they managed to spontaneously bring King to speak publicly. After speaking at the Marienkirche to a crowd of roughly 3,000 people, King then appeared at the Sophienkirche later in the afternoon in order that others who were not able to hear his first speech could attend. The overwhelming response from the East Berliners left a deep impression upon King, who spoke openly against discrimination and condemned the Berlin Wall. "Here are God's children on both sides of the wall, and no man-made barrier can destroy this fact. With this faith we will be able to tear out of the mountain of despair a stone of hope." See Lars-Broder Keil, "When Martin Luther King Jr. Spoke to East Berlin," *ozy.com*, January 18, 2016. www.ozy.com/flashback/when-martin-luther-king-jr-spoke-to-east-berlin/34061 (accessed May 2017).

26 The idea of the "Church for Others" is drawn from the writings of Dietrich Bonhoeffer, whose theological teachings were deeply influential in the context of the German churches. As Bonhoeffer wrote in his *Letters and Papers from Prison*: "The Church is the Church only when it exists for others ... not dominating, but helping and serving. It must tell men of every calling what it means to live for Christ, to exist for others." See Dietrich Bonhoeffer, *Letters and Papers from Prison* (London: SCM Press, 2001), 33.

27 Sabrina P. Ramet, *Nihil Obstat: Religion, Politics, and Social Change in East-Central Europe and Russia* (Durham, NC: Duke University Press, 1998), 79.

28 Ibid.

29 Christian Führer, quoted in Wayne C. Bartee, *A Time to Speak Out: The Leipzig Citizen Protests and the Fall of East Germany* (Westport, CT: Praeger Publishers, 2000), 169.

30 Václav Havel, "The Power of the Powerless," in Václav Havel et al., *The Power of the Powerless: Citizens Against the State in Central-Eastern Europe*, ed. John Keane (Armonk, NY: M. E. Sharpe, 1990), 36–37.

31 Ibid., 30.

32 Václav Benda, *The Parallel Polis* (Amsterdam: Second Culture Press, 2015, pamphlet).

33 Havel, *The Power of the Powerless*, 95.

34 In response to a crackdown by the police in late 1976, Charter 77 was an informal document criticizing the government for human rights abuses and was widely circulated in early 1977 through clandestine publications. The Charter acted as an important guide throughout the Soviet occupation, and following the Velvet Revolution in 1989 many of its signatories and central members were active in negotiating the transfer of power and establishing a new government.

35 The "solidarity of the shaken" was a term coined by the philosopher Jan Patočka to describe the culture of the underground. Quoted in Václav Havel, *Open Letters: Selected Writings 1965–1990* (New York: Vintage Books, 1992), 271.

36 In conversation with the author, September 2016, Berlin.

37 Susan Bickford, *The Dissonance of Democracy: Listening, Conflict, and Citizenship* (Ithaca, NY: Cornell University Press, 1996), 97.

38 Ibid., 162.

39 Ibid., 129.

40 See Nick Couldry, *Why Voice Matters: Culture and Politics after Neoliberalism* (London: Sage Publications, 2010).

41 Havel, *The Power of the Powerless*, 79.

42 See Joan Walsh, "The Abu Ghraib Files," *Salon.com*, March 14, 2006. www.salon.com/2006/03/14/introduction_2/ (accessed February 2017.)

43 Judith Butler, *Frames of War: When is Life Grievable?* (London: Verso, 2016), xvi.

44 Ibid., 61.

45 Ibid., 23.

46 James C. Scott, *The Weapons of the Weak: Everyday Forms of Peasant Resistance* (New Haven: Yale University Press, 1985).

47 See Asef Bayat, "Revolution without Movement, Movement without Revolution: Comparing Islamic Activism in Iran and Egypt." *Comparative Studies in Society and History* 40(1) (January 1998): 136–69.

48 Errol Babacan, "The Gezi Commune: Experiences of an Uprising." *Krytyka: Global Activism*, Special Issue (2014): 21.

6

Poor Acoustics: Listening from Below

In developing the concept of sonic agency, I've attempted to elaborate how it may come to act in relation to contemporary social and political struggles. This is furthered by considering how people draw from experiences of sound and listening in order to join together against conditions of loss and powerlessness. From such work, the figures of invisibility, overhearing, the itinerant, and the weak are posited as the basis for specific emancipatory practices. In particular, I've sought to mobilize these figures to lead us toward a rethinking of the political as being solely an arena of visibility and open public discourse. My attention is drawn toward individuals and communities that pry open agentive possibilities often through an art of trespass and survival, withdrawal and commoning that reconstitutes the political realm and what counts as appearance.

Through invisibility and overhearing I attempt to map the everyday yet complex ethics brought forward from encounters with strangers and the strange, the disappeared and the hidden. This leads to reflections on particular political struggles and how the conditions of network culture support as well as corral the potentials of diverse assemblages. The invisible and interruptive technologies of surveillance and the related economies of attention today, for example, reform subjectivity as *unhomed*, captured and integrated into assemblages that are more vibratory than pictorial, more overheard than heard. Within such conditions, a distributed sense of agency emerges to occupy global networks through "stranger relations," leading to a new ontology of the social in which listening and being heard contend with an intensity of noise, an otherness always closer than imagined. From this new condition, which Jane Bennett hints at by way of her concept "vibrant matter," practices of commoning, collective making, and creative instituting emerge, which work through the contemporary logics of immaterial labor, transnational culture, and precarity to spirit new formations of social power and civic responsibility. Moving from conditions of network life, as one aligned with strangerhood and the potential of the overheard, I examine the topic of the disappeared, particularly in the context of Chile. The disappeared forces us into spectral territories of invisibility and the missing, which require another set of discursive and perceptual tools. Against and alongside the disappeared, I put forward the acousmatic as a powerful device or construct, one that specifically affords a route for contending with absence. The invisible quality of sounds that we hear without seeing their source (often utilized in cinema and electro-acoustic music practices) provides the basis for probing how subjectivity is deeply entangled with visibility and the work of appearance. In contrast, the disappeared and the invisible demand a shift in understanding, as well as practices of communication and remembering, that give way to a potentiality found in the dark. I've attempted to pose invisibility as the basis for reorienting subjectivity as being founded on appearance and the directives of the clearly outlined; instead, liminalities and shadowed realities lead us into relations with the unnamable and the unwanted: the disappeared whose silences and forced withdrawal pressure the seen. Invisibility and overhearing are scenes of relational figuring that extend the conditions of appearance toward peripheries where face-to-face encounters are complicated by the erased and the ghostly, as well as by hidden agents of control. Through problematizing and haunting the space of appearance, and what counts as a subject proper, and by interrupting the built environment and ourselves with strange sounds, invisibility and overhearing turn us toward the potential of acousmatic constructs and the interruptive operations of noise, this unfamiliar or muted voice that requests of me a greater consideration.

In addition, with the itinerant and the weak we have been led into reflections on poetic knowledge and the collective reverberations by which speaking and acting extend their agentive reach. Whether in the context of the post-colonial Antilles, or the former East German state, emancipatory struggles often find support through affective and lyrical practices, the creative modulations of shouting, singing, and speaking up, and embodied expressions of moral compassion. In particular, I've sought to consider experiences and conditions of itinerancy and weakness as the basis from which new solidarities and constructs of collective dissent or escape are produced. By examining the loving ethos of the counter-culture in the late 1960s, and the struggles of East German peace movements, an understanding is opened up as to the importance moral orientation and conscientious objection may play in forming cultures of civic responsibility. Here, weakness is repositioned as an expression of moral strength, one deeply aligned with the silences of refusal and the reverberations of hope found in standing still and sitting in, as well as through instituting forms of common living. As Václav Havel argues, the "power of the powerless" may be found in the moral belief in a future to come and which is given support through acting "in solidarity" with the hidden spheres of "the shaken."[1] In addition, I've been interested to consider how itinerancy is repositioned so as to give shelter against the shattering experiences of eviction and expulsion prevalent today, as well as systems of colonial occupation. From within the conditions of the expelled or the exiled are crafted a range of practices by which resistances and self-determination are nurtured and deployed, often through the appropriation and modulation - a critical echoing - of dominant languages and the lyrical drive of poetics (as in the example of what Aimé Césaire would term "black French"). Loving ethos, diasporic echoes, non-violent resistances - within these fragile positions we find not only voices lifted up, but also powerful acts that sound out an acoustical force of social becoming.

In focusing on emancipatory practices through sonic agency, which is ultimately a set of narratives about emergent subjectivities - a queering of the limits that define bodies out of place - sound is captured as a means by which to animate forms of resistance and unlikely publics. In this context, sonic discourses are applied to critical understandings of public life, political subjectivity, dissident cultures, and in support of an ethics for the transnational present by aligning us with the particular epistemologies and ontologies, emotionalities and imaginaries drawn from sounded experience and audition. From my perspective, the potentiality of what Arendt terms "speech and action" undertaken together, in constituting the formation of political space and process, gains traction through an expanded reflection upon the degree to which audition and the expressivities promulgated by sounded experience become central to working out the concerns of the world. Speech and action are subsequently cast not only as openly verbal and visual, but equally as forces of vibrancy, shaped and reshaped through the tonalities and resonances passing through bodies and places, inflected and flexed by modulations of silence and the volumes always punctuating acts of joining together.

In doing so, I'm interested to support the speech and actions that turn us toward not only those we may see, according to particular codings and decodings, or even those that may assemble in a public space, but equally those unlikely publics often taking shape from within the entangled depths of life; unhomed publics whose tactics unfold other pathways and powers drawn from the knowledges of displaced bodies and floating subjects, and the sensuous materiality of creative and critical ideas. These publics gain momentum in the world by nurturing a range of capacities and resources, some of which I work to outline: the invisible, the overheard, the itinerant, and the weak are highlighted as modalities by which subjects contend with dominant structures and assemblies saturated with imbalances of power. As such, they interrupt constructs of the public sphere as being fundamentally constituted through the capacities of discourse and visibility.

In *Crisis to Insurrection*, Mikkel Bolt Rasmussen considers the wave of protests occurring throughout 2011–12 as expressive of not only a general attack on capitalism and the neoliberal city, but equally of a new political subject grounded in the transnational present. Within the squares "a new political subject" emerged by "refusing the established order and using the city to do something different from just consuming or working."[2] The gestures of joining together and of occupying spaces, of placing oneself as an obstacle to the continual dominance of financial operations, for Rasmussen, underscores the mobilization of this new political subject, as one "refusing and showing this refusal."[3] Importantly, the ways in which Occupy Wall Street, for example, was characterized by a wide range of seemingly incongruous demands or issues, or by "failing" to proceed according to conventions of oppositional politics or by way of a social movement with particular leadership, demonstrates "an attempt to produce a political space that precedes the daily political procedures whereby all places are always already distributed."[4] Instead, Occupy Wall Street, along with other occupations and protests, demanded a "redemocratization" of social and political systems and spaces.

Within Zuccotti Park, as well as Syntagma Square, the Puerta del Sol, and in Tahrir Square, redemocratization became a governing principle and organizational directive, guiding particular discussions and initiatives, as well as the micropolitics and social exchanges internal to the square. Consequently, the new spirit of political subjectivity finds momentum through the emergent cultures of self-organization and participatory work today, which support civic generosity, nonmovements and quotidian encroachments, deviating and inter-lingual voices, and practices of radical care.

While references to Occupy, the Movement of the Squares, and the Arab Spring, for example, remain poignant and relevant, the more recent intensification of populism, far-right movements, and illiberalism continues to fragment the political establishment as well as the general fabric of social relations, inciting and supporting racism, sexism, religious intolerance, and a deepening of nationalistic sentiment, all of which give further challenge to the movements and expressions of social solidarity. The post-factual project of Trump, underscored by advisor Kellyanne Conway's quip about "alternative facts" in support of statements from the White House, and the dissemination of "fake news" that continues to spread throughout various media platforms, are but additional signs of an overall moral deterioration of public offices. The complex emergence of illiberal agendas and projects is sparking equally tensed responses; new movements of resistance and social solidarity appear across Europe and North America aimed at countering the ugly face of "new fascism." These more "liberal" movements and formations may be understood to express the "new political subjectivity" Rasmussen highlights, yet they are in turn mirrored by the new political subject of the alt-right that bolsters the politics of Trump, whose success has been predicated partly upon casting himself as a "political outsider" and allied with the downtrodden "white poor." (The result of the Brexit vote is also partially founded upon similar political subjects, their grievances as well as hopes.)

The rhetoric of "revolution" espoused by Trump, Nigel Farage or Marine Le Pen against the political establishment, and in support of common or disaffected people, forces pause when considering new political subjectivity aimed at the redemocratization of the system. What might the complex emergence of new political subjects on both the liberal and illiberal ends of the spectrum suggest or indicate in terms of defining political work today? How might we identify, from within the new features of protest culture and emancipatory practices, pathways toward a greater renewal of social solidarity aimed at nurturing life with others?

Returning to Balibar's notion of insurrection as being fundamental to democracy, is not the drive of an insurrectionary sensibility a type of desire set in motion by visions of possible worlds? An imaginary that takes its guidance through creative expressions and the wish to live within meaningful conditions? An urgency that works to intensify direct relations and to remake the constructs of daily life? The illiberal movements today must therefore be considered as being based on a parallel drive to "redemocratize" the system – for instance, within Europe, where the European Union has mostly failed to democratically foster a "European citizenry." The Brexit vote and the Trump election can be seen to equally capture a basic wish on the part of some to reconfigure power relations today. Critical distinctions therefore need to be made. What particular world-view drives the project of redemocratization within specific contexts? What civic structures and pedagogical platforms need to be cultivated to nurture political subjectivity as it relates to the complexity of a global reality? And how may this subject work to reshape the project of "revolution" promulgated by illiberalism toward less nationalistic and racist ends?

The "wrench of equality" Jacques Rancière highlights, in underpinning politics, is one that arises from those without; it is necessarily a force coming up from below, from edges and peripheries, or from *within*, through the floating networks or institutional frameworks, the subjects and bodies that force their way in or that exit the scene – *life in the making* is always a search for possibilities. The urgency of this equality comes to interrupt the acoustics of assembly with expressions of the dispossessed and disenfranchised, the desire and the longing of those out of place; instead, a *poor acoustics* defines the vitality of unlikely publics, to reorient the demarcations of the heard and the underheard, and the properly sounded. How to bring forward a critical listening within the increasingly complex arena of post-factual realities? From within the poor acoustics of the underheard and the unlikely, how might we nurture a particular aural attention in order to steer the project of redemocratization toward understandings of global responsibility and compassion? To craft from the dissonances and consonances central to political life the basis for new states of social solidarity and imagination? A poor acoustic by which to tense the smooth delivery of lies appearing as facts?

For what may constitute the critical difference within today's "revolutionary" atmosphere is whether one is able to navigate the post-factual and the illiberal in support of deep knowing and mutual recognition: the giving toward a future whose "wrench of equality" may be wielded in support of far-reaching civic mindfulness. As Jamie Heckert eloquently states, the ethics required to create more egalitarian systems is founded on "the dignity of listening to oneself, and the dignity of being listened to."[5]

From such perspectives and concerns, I am led to the question of listening acting as a potential form of interruption. If, as I've tried to outline, sound and listening are deeply enabling for social exchange, acting as essential means for encountering and sharing differences while nurturing mutual recognition, can we envision a broader project? A type of amplification of this listening that may cater to a multiplicity of narratives and accounts, stories and their lessons? Is it possible to cast listening as an activism that may give challenge to existing demarcations or structures of domination, or against those that seek to dominate others? As Kate Lacey suggests, what is needed within today's environment is not only to secure the right to free speech, but also, to turn our attention to freedom of listening. Freedom of listening is posited as being essential for enabling a "plurality" of voices to be heard; in other words, freedom of listening produces an extremely active relational space within which voices may resound. Yet, the potentiality of freedom of listening may aid in discovering and nurturing new formations of solidarity by also explicitly relating us to things *beyond* the voice. The silences of still bodies, the vibrational and rhythmical intensities of collective acts, the tonalities disturbed or distributed by cacophonic volumes, and the co-soundings and echoes of earthly creatures and matters – these are equally defining of the public sphere and expressions of political desire. To enact one's freedom of listening is to necessarily aim for a broader and richer engagement with the range of voices and things to be heard and shared.

Listening activism, as I begin to understand it, gives elaboration to the many forms listening may take; deep as well as shallow listening, horizontal and vertical hearings; a listening that flexes itself, that surrenders as well as punctuates: a listening around or through, toward or against others – *this listening that I give as well as through which I take.* Forms of listening are ultimately productions of subjects and sites, knowledges and relations, contouring and shaping the subjective and the intersubjective, the energetic and the material features that greatly affect personal and political life. Listening is often tuning us to the interplay of meaningful layers that constitute the world, bridging the seen and the unseen, foreground and background, things and bodies with animate forcefulness: listening draws one in, toward certain depths, while drawing out the underheard into greater volume – a poor acoustic whose dirtiness is reflective of the tussle that is public life.

What may happen then in instances of collective listening? A listening together taking place not only within spaces of music, for instance, but drawn out into the open? That may move us, as music often does, into states not only of euphoric dance, but also into other types of movement? This might be a listening activism directed at particular sites, for instance, around situations of conflict or within communities, applied to spaces of presence or emptiness, locating us around that which is missing. In these situations, listening may express concern as well as indignation by bringing attention to the said and unsaid – *this sound that relates us to the not-yet-apparent.* A gathering of listeners, in the squares, or in the classrooms and market places, the backrooms and storefronts, may perform to create a gap, a duration drawn out, detouring the flows of normative actions, of declarations and decrees, with a persistent intensity – a nagging quietude, possibly: this act of *doing listening,* together; and by gathering attention it may also create an image: the image of the listener as one who *enacts* attention or consideration and, in doing so, nurtures the conditions for mindful engagement. The listening ... that works ... that interrupts, or ... that beats back ... to produce ... in those gaps of ... time and space ... another pause: the ... interval in which ... something ... someone, or ... ... others ... force onto particular contexts ... – the classroom, the hall of justice, the park ... or the home – this attention ... the listening that ... ... demands ... ... and that gives ... and that may resonate ... that may amplify ... the potentiality of ... being ... side by side ...

Freedom of listening, as I'm describing, sets the conditions for dwelling within the present with others – for this sound we hear is already the production of a shared world, however tensed or disjunctive, this sound that animates a space between, and that is always moving on and through and with. In doing so, listening is the expression of an "art of presence," crafting from the body and its place in the world and with others new formations of social becoming.[6]

If sound is a force that continually stirs the surroundings, driving forward an array of vibrations and reverberations, echoing across borders and rippling relations between interiorities and exteriorities, the inner depths with surfaces, and beyond, collecting singularities into a collective body, or forcing them apart, it does so in such a way as to *potentialize* the inherent flux of things – to intensify the animate conditions of life in the making. The agitating and evanescent project of sound is fundamentally a *disquieting* presence, one nestled within the stability of the dominant and which may come to life under the force of a sudden breath, or a surprising voice, to counter the demarcations of the visible, of who or what counts, through practices of the invisible and the not-yet-apparent; to interrupt the technocratic enclosures of the commons by extending assemblages of social vitality through the overheard and the strangers found therein; to give challenge to all types of borders by supporting the migrations of the dispossessed, the floating, the *echo-subjects,* and those always already in search; and to nurture the means to stage our weaknesses as the basis for a greater strength, the strength found in erotic knowledge and shared vulnerability. These are positions and practices, capacities and imaginaries given traction by the freedom of listening, by listening to oneself in order to deepen one's conscience and consciousness, and from which to hear others, as they resound with particular indignation or hope. From such instances, one may begin to truly sense the interdependencies of which one is always a part, and which may encourage a collective making of this life lived ...

Notes

1 See Václav Havel, "The Power of the Powerless," in Václav Havel et al., *The Power of the Powerless: Citizens Against the State in Central-Eastern Europe*, ed. John Keane (Armonk, NY: M. E. Sharpe, 1990), 42.

2 Mikkel Bolt Rasmussen, *Crisis to Insurrection: Notes on the Ongoing Collapse* (Wivenhoe: Minor Compositions, 2015), 79.

3 Ibid.

4 Ibid.
5 Jamie Heckert, "Listening, Caring, Becoming: Anarchism As an Ethics of Direct Relationships," in B. Franks and M. Wilson (eds.) *Anarchism and Moral Philosophy* (Basingstoke: Palgrave Macmillan, 2010), 186–207.
6 See Asef Bayat, "Revolution without Movement, Movement without Revolution: Comparing Islamic Activism in Iran and Egypt." *Comparative Studies in Society and History* 40(1) (January 1998): 136–69.

Bibliography

Agamben, Giorgio. *Homo Sacer: Sovereign Power and Bare Life*. Stanford: Stanford University Press, 1998.

Anzaldúa, Gloria. "La conciencia de la mestiza: Towards a New Consciousness," in Gloria Anzaldúa (ed.), *Making Face, Making Soul: Haciendo Caras. Creative and Critical Perspectives by Women of Color.* San Francisco: Aunt Lute Foundation Books, 1990: 377–89.

Anzaldúa, Gloria. "Haciendo caras, una entrada," in Gloria Anzaldúa (ed.), *Making Face, Making Soul: Haciendo Caras. Creative and Critical Perspectives by Women of Color.* San Francisco: Aunt Lute Foundation Books, 1990: xv–xxviii.

Anzieu, Didier. *The Skin-Ego*. London: Karnac Books, 2016.

Appadurai, Arjun. *Modernity at Large: Cultural Dimensions of Globalization.* Minneapolis: University of Minnesota Press, 1996.

Appiah, Kwame Anthony. *Cosmopolitanism: Ethics in a World of Strangers.* London: Penguin Books, 2006.

Applebaum, Anne. *Iron Curtain: The Crushing of Eastern Europe 1944–1956.* London: Penguin Books, 2012.

Arendt, Hannah. *The Human Condition*. Chicago: University of Chicago Press, 1998.

Ataç, Ilker, Köster-Eiserfunke, Anna, and Schwiertz, Helge. "Governing through Citizenship and Citizenship from Below: An Interview with Kim Rygiel." *movements. Journal für kritische Migrations- und Grenzregimeforschung* 1(2), 2015. http://movements-journal.org/issues/02.kaempfe/02.rygiel,ataç,köster-eiserfunke,schwiertz--governing-citizenship-from-below.html (accessed March 2017).

Attali, Jacques. *Noise: Political Economy of Music*. Minneapolis: University of Minnesota Press, 1988.

Babacan, Errol. "The Gezi Commune: Experiences of an Uprising." *Krytyka: Global Activism*, Special Issue, 2014: 19–21.

Bhabha, Homi K. *The Location of Culture*. London: Routledge, 1994.

Balibar, Étienne. *Citizenship*. Cambridge: Polity Press, 2015.

Baudrillard, Jean. *For a Critique of the Political Economy of the Sign*. Candor, NY: Telos Press Publishing, 1981.

Bauman, Zygmunt. *Strangers at Our Door*. Cambridge: Polity Press, 2016.

Bayat, Asef. *Life as Politics: How Ordinary People Change the Middle East.* Amsterdam: Amsterdam University Press, 2010.

Bayat, Asef. "Revolution without Movement, Movement without Revolution: Comparing Islamic Activism in Iran and Egypt." *Comparative Studies in Society and History* 40(1), January 1998: 136–69.

Benda, Václav. *The Parallel Polis*. Amsterdam: Second Culture Press, 2015, pamphlet.

Bennett, Jane. *The Enchantment of Modern Life: Attachments, Crossings, and Ethics*. Princeton: Princeton University Press, 2001.

Bennett, Jane. *Vibrant Matter: A Political Ecology of Things*. Durham, NC: Duke University Press, 2010.

Berardi, Franco Bifo. *The Soul at Work*. Los Angeles: Semiotext(e), 2009.

Berardi, Franco Bifo. *The Uprising: On Poetry and Finance*. Los Angeles: Semiotext(e), 2012.

Bernabé, Jean, Chamoiseau, Patrick, and Confiant, Raphaël. "In Praise of Creoleness." *Callaloo* 13(4), Autumn 1990: 886–909.

Bartee, Wayne C. *A Time to Speak Out: The Leipzig Citizen Protests and the Fall of East Germany*. Westport, CT: Praeger Publishers, 2000.

Bickford, Susan. *The Dissonance of Democracy: Listening, Conflict, and Citizenship*. Ithaca, NY: Cornell University Press, 1996.

Bonhoeffer, Dietrich. *Letters and Papers from Prison*. London: SCM Press, 2001.

Brathwaite, Kamau. *The Development of Creole Society in Jamaica, 1770–1820*. Kingston and Miami: Ian Randle Publishers, 2005.

Breton, André. *Martinique: Snake Charmer*. Austin: University of Texas Press, 2008.

Brown, Wendy. "We Are All Democrats Now ...." In *Democracy in What State?* New York: Columbia University Press, 2012: 44–57.

Burkart, Patrick. *Pirate Politics: The New Information Policy Contests*. Cambridge, MA: MIT Press, 2014.

Butler, Judith. "Bodies in Alliance and the Politics of the Street." Lecture given in Venice, September 7, 2011, in the framework of the series The State of Things, organized by the Office for Contemporary Art Norway (OCA). www.eipcp.net/transversal/1011/butler/en (accessed June 2015).

Butler, Judith. *Frames of War: When is Life Grievable?* London: Verso, 2016.

Butler, Judith. *Notes toward a Performative Theory of Assembly*. Cambridge, MA: Harvard University Press, 2015.

Butler, Judith. *Precarious Life: The Powers of Mourning and Violence*. London: Verso, 2006.

Césaire, Aimé. *Notebook of a Return to My Native Land*. Hexham, Northumberland: Bloodaxe Books, 2010.

Césaire, Suzanne. "A Civilization's Discontent." In Michael Richardson (ed.), *Refusal of the Shadow: Surrealism and the Caribbean*. London: Verso, 1996: 96–100.

Chion, Michel. *Audio-Vision: Sound on Screen*. New York: Columbia University Press, 1994.

Chion, Michel. *The Voice in Cinema*. New York: Columbia University Press, 1999.

Clarke, Peter B. *Black Paradise: The Rastafarian Movement*. San Bernardino, CA: The Borgo Press, 1994.

Clausen, Marco. "Prinzessinnengarten & The Neighborhood Academy." *Free Berlin* 3, June 2016: 22–27.

Cohen, Robin and Sheringham, Olivia. *Encountering Difference: Diasporic Traces, Creolizing Spaces*. Cambridge: Polity Press, 2016.

Conio, Andrew. "Revolutionary: What Do You Want?" In *The Living School*. London: South London Gallery, forthcoming.

Couldry, Nick. *Why Voice Matters: Culture and Politics after Neoliberalism*. London: Sage Publications, 2010.

Davis, Gregson. "Forging a Caribbean Literary Style: 'Vulgar Experience' and the Languages of Césaire's *Cahier d'un retour au pays natal*." *South Atlantic Quarterly* 115(3), July 2016: 457–68.

Dewey, Fred. *The School of Public Life*. Berlin: Errant Bodies Press, 2015.

Döşemeci, Mehmet. "Don't Move, Occupy! Social Movement vs. Social Arrest." *Krytyka: Global Activism*, Special Issue, 2014: 10–17.

Dyson, Frances. *The Tone of Our Times: Sound, Sense, Economy, and Ecology*. Cambridge, MA: MIT Press, 2014.

Ellison, Ralph. *The Invisible Man*. New York: Vintage Books, 1995.

Ezcurra, Mara Polgovsky. "Zona De Dolor: Body and Mysticism in Diamela Eltit's Video-Performance Art." *Journal of Latin American Cultural Studies: Travesia* 21(4), 2012: 517–33.

Flusser, Vilém. *The Freedom of the Migrant: Objections to Nationalism*. Urbana: University of Illinois Press, 2003.

Fraser, Nancy. "Rethinking the Public Sphere: A Contribution to the Critique of Actually Existing Democracy." *Social Text* 25/26, 1990: 56–80.

Freire, Paulo. *Education for Critical Consciousness*. London: Bloomsbury Academic, 2013.

Funder, Anna. *Stasiland: Stories from Behind the Berlin Wall*. London: Granta Publications, 2011.

García-Saavedra, Soledad. "Secret Archives: Suspension and Overflow in The Library of No-History." In Cristián Gómez-Moya (ed.), *Human Rights, Copy Rights: Visual Archives in the Age of Declassification*. Santiago: Museum of Contemporary Art/University of Chile, 2013: 93–108.

García-Saavedra, Soledad. "Thunder Stealers: From Phantasmatic Apparitions to Ghostly Bodies in Chile." In *The Invisible Seminar*. Bergen: Institute of Art, University of Bergen, forthcoming.

Geheimagentur, Schäfer, Martin Jörg, and Tsianos, Vassilis S. (eds.). *The Art of Being Many: Towards a New Theory and Practice of Gathering*. Bielefeld: transcript Verlag, 2016.

Gilroy, Paul. *The Black Atlantic: Modernity and Double-Consciousness*. Cambridge, MA: Harvard University Press, 1995.

Ginsburg, Faye. "Rethinking the Digital Age." In Pamela Wilson and Michelle Stewart (eds.), *Global Indigenous Media: Cultures, Poetics, and Politics*. Durham: Duke University Press, 2008: 287–305.

Glissant, Édouard. *Poetics of Relation*. Ann Arbor: University of Michigan Press, 2010.

Gordon, Avery F. *Ghostly Matters: Haunting and the Sociological Imagination*. Minneapolis: University of Minnesota Press, 2008.

Graeber, David. *Fragments of an Anarchist Anthropology*. Chicago: Prickly Paradigm Press, 2004.

Habermas, Jürgen. *The Structural Transformation of the Public Sphere: An Inquiry into a Category of Bourgeois Society*. Cambridge: Polity Press, 2015.

Harvey, Stefano and Moten, Fred. *The Undercommons: Fugitive Planning and Black Study*. New York: Minor Compositions, 2013.

Havel, Václav. *Open Letters: Selected Writings 1965–1990*. New York: Vintage Books, 1992.

Havel, Václav. "The Power of the Powerless." In Václav Havel et al., *The Power of the Powerless: Citizens Against the State in Central-Eastern Europe*, ed. John Keane. Armonk, NY: M. E. Sharpe, 1990: 23–96.

Hebdige, Dick. *Cut 'n' Mix: Culture, Identity and Caribbean Music*. London: Routledge, 1987.

Heckert, Jamie. "Listening, Caring, Becoming: Anarchism As an Ethics of Direct Relationships." In B. Franks and M. Wilson (eds.), *Anarchism and Moral Philosophy*. Basingstoke: Palgrave Macmillan, 2010: 186–207.

Henriques, Julian. *Sonic Bodies: Reggae Sound Systems, Performance Techniques, and Ways of Knowing*. New York: Continuum, 2011.

Hoffman (Free), Abbie. *Revolution for the Hell of It*. New York: Dial Press, 1968.

hooks, bell. *Outlaw Culture: Resisting Representations*. New York: Routledge, 1994.

hooks, bell. *Salvation: Black People and Love*. New York: HarperCollins, 2001.

hooks, bell. *Teaching to Transgress: Education as the Practice of Freedom*. New York: Routledge, 1994.

Huxley, Aldous. *The Doors of Perception*. London: Vintage, 2004.

Jirous, Ivan. *A Report on the Third Czech Musical Revival*. Amsterdam: Second Culture Press, 2015 (pamphlet, originally pub. 1975).

Keil, Lars-Broder. "When Martin Luther King Jr. Spoke to East Berlin." In *ozy.com*, January 18, 2016. www.ozy.com/flashback/when-martin-luther-king-jr-spoke-to-east-berlin/34061 (accessed May 2017).

Lacey, Kate. *Listening Publics: The Politics and Experience of Listening in the Media Age*. Cambridge: Polity Press, 2013.

Leary, Timothy with Metzner, Ralph and Alpert, Richard. *The Psychedelic Experience*. London: Penguin Books, 2008.

Levinas, Emmanuel. *Totality and Infinity: An Essay on Exteriority*. Pittsburgh: Duquesne University Press, 1998.

Lingis, Alphonso. *The Community of Those Who Have Nothing in Common*. Bloomington: Indiana University Press, 1994.

Lorde, Audre. "Uses of the Erotic: The Erotic as Power." In *Sister Outsider: Essays and Speeches*. Berkeley: Crossing Press, 2007: 53–59.

Lorey, Isabell. *State of Insecurity: Government of the Precarious*. London: Verso, 2015.

McLuhan, Marshall. *Understanding Media: The Extensions of Man*. London: Routledge, 2002.

Mailer, Norman. *The Armies of the Night: History as a Novel, The Novel as History*. New York: Penguin, 1968.

Marazzi, Christian. *The Violence of Financial Capital*. Los Angeles: Semiotext(e), 2010.

Marcos, Subcomandante Insurgente. *Our Word is Our Weapon: Selected Writings*. London: Serpent's Tail, 2001.

Marcuse, Herbert. *One-Dimensional Man: Studies in the Ideology of Advanced Industrial Society*. Boston: Beacon Press, 1964.

Merleau-Ponty, Maurice. *The Visible and the Invisible*. Evanston, IL: Northwestern University Press, 1968.

Miranda, Maria. *Unsitely Aesthetics: Uncertain Practices in Contemporary Art*. Berlin: Errant Bodies Press, 2013.

Morales, Evo. International Summit against Imperialism, 2013, in which Morales announced a six-point strategy for sovereignty. https://libya360.wordpress.com/2013/09/27/bolivia-against-colonialism-and-imperialism-six-strategies-for-sovereignty-dignity-and-the-life-of-the-peoples/ (accessed December 2016).

Moten, Fred. *In the Break: The Aesthetics of the Black Radical Tradition*. Minneapolis: University of Minnesota Press, 2003.

Mouffe, Chantal. *The Democratic Paradox*. London: Verso, 2005.

Neumark, Norie. *Voicetracks: Attuning to Voice in Media and the Arts*. Cambridge, MA: MIT Press, 2017.

Ramet, Sabrina P. *Nihil Obstat: Religion, Politics, and Social Change in East-Central Europe and Russia*. Durham, NC: Duke University Press, 1998.

Rancière, Jacques. "Democracies against Democracy." In *Democracy in What State?* New York: Columbia University Press, 2012: 76–81.

Rancière, Jacques. *Disagreement: Politics and Philosophy*. Minneapolis: University of Minnesota Press, 1999.

Rancière, Jacques. *On the Shores of Politics*. London: Verso, 2007.

Rasmussen, Mikkel Bolt. *Crisis to Insurrection: Notes on the Ongoing Collapse*. Wivenhoe: Minor Compositions, 2015.

Richard, Nelly. "Margins and Institutions: Art in Chile since 1973." *Art and Text* 21, Special Issue, May–July 1980.

Ricoeur, Paul. *On Translation*. New York: Routledge, 2006.

Rolnik, Suely. "Anthropophagic Subjectivity." In *Arte Contemporânea Brasileira: Um e/entre Outro/s*. São Paulo: Fundação Bienal de São Paulo, 1998: 1–34. https://whcavantgardes.files.wordpress.com/2014/09/anthropophagic-subjectivity.pdf (accessed February 2017).

Roszak, Theodore. *The Making of a Counter Culture: Reflections on the Technocratic Society and Its Youthful Opposition*. Garden City, NY: Anchor Books, 1969.

Rygiel, Kim. *Globalizing Citizenship*. Vancouver: University of British Columbia Press, 2011.

Sarachild, Kathie. "Consciousness-Raising: A Radical Weapon." Talk given at the First National Conference of Stewardesses for Women's Rights in New York City, March 12, 1973. https://organizingforwomensliberation.wordpress.com/2012/09/25/consciousness-raising-a-radical-weapon/ (accessed October 2016).

Sassen, Saskia. *Expulsions: Brutality and Complexity in the Global Economy*. Cambridge, MA: The Belknap Press of Harvard University Press, 2014.

Scott, James C. *The Weapons of the Weak: Everyday Forms of Peasant Resistance*. New Haven: Yale University Press, 1985.

Seligmann, Katerina Gonzalez. "Poetic Productions of Cultural Combat in *Tropiques*." *South Atlantic Quarterly* 115(3), July 2016: 495–512.

Sennett, Richard. *The Uses of Disorder: Personal Identity and City Life*. New Haven: Yale University Press, 2008.

Silverman, Kaja. *The Acoustic Mirror: The Female Voice in Psychoanalysis and Cinema*. Bloomington: Indiana University Press, 1988.

Simmel, Georg. "The Stranger." In *On Individuality and Social Forms: Selected Writings*. Chicago: University of Chicago Press, 1971.

Simone, AbdouMaliq. Informal talk at the ifa gallery, Berlin, 2016.

Simone, AbdouMaliq. "People as Infrastructure: Intersecting Fragments in Johannesburg." *Public Culture* 16(3), 2004: 407–29.

Sitrin, Marina and Azzellini, Dario. *They Can't Represent Us! Reinventing Democracy from Greece to Occupy*. London: Verso, 2014.

Sparrow (Slinger Francisco). "Dan is the Man." In Stewart Brown, Mervyn Morris, and Gordon Rohlehr (eds.), *Voiceprint: An Anthology of Oral and Related Poetry from the Caribbean*, Harlow: Longman, 1989: 130.

Stalder, Felix. *Digital Solidarity*. London: Mute Books, 2013.

Standing, Guy. *The Precariat: The New Dangerous Class*. New York: Bloomsbury Academic, 2014.

Stavrides, Stavros. *Common Space: The City as Commons*. London: Zed Books, 2016.

Steen, Bart van der, Katzeff, Ask, and Hoogenhuijze, Leendert van (eds.). *The City is Ours: Squatting and Autonomous Movements in Europe from the 1970s to the Present*. Oakland, CA: PM Press, 2014.

The Struggles Collective. "Lessons from the Struggles: A Collage." *movements. Journal für kritische Migrations- und Grenzregimeforschung* 1(2), 2015. http://movements-journal.org/issues/01.grenzregime/19.from-the-struggles--lessons.html (accessed March 2017).

Sundaram, Ravi. *Pirate Modernity: Delhi's Media Urbanism*. London: Routledge, 2010.

Taussig, Michael. *The Magic of the State*. New York: Routledge, 1997.

Terranova, Tiziana. *Network Culture: Politics for the Information Age*. London: Pluto Press, 2004.

Tsianos, Vassilis S. and Tsomou, Margarita. "Assembling Bodies in New Ecologies of Existence: The Real Democracy Experience as Politics Beyond Representation." In Geheimagentur, Martin Jörg Schäfer, and Vassilis S. Tsianos (eds.), *The Art of Being Many: Towards a New Theory and Practice of Gathering*. Bielefeld: transcript Verlag, 2016: 77–94.

Tucker, Irwin St. John. "Forward the Hobo! A Plea for a National Service Army" (undated), from the archives at the University of Chicago.

Ultra-red. "Dundee [2011]." From the self-published notebook series, *Nine Workbooks, 2010–2014*. www.ultrared.org.

Ultra-red. "In/visibility and the Conditions of Collective Listening." In Brandon LaBelle (ed.), *The Invisible Seminar*. Bergen: Institute of Art, University of Bergen, forthcoming.

Ünsal, Nadiye. "Challenging 'Refugees' and 'Supporters': Intersectional Power Structures in the Refugee Movement in Berlin." *movements. Journal für kritische Migrations- und Grenzregimeforschung* 1(2), 2015. http://movements-journal.org/issues/02.kaempfe/09.ünsal--refugees-supporters-oplatz-intersectionality.html (accessed February 2017).

U.S. Covert Actions by the Central Intelligence Agency in Chile. The Church Committee Report and the Hinchey Report as Presented to the US Congress. Rockville, MD: Arc Manor, 2008.

Valdés, Adriana. "Voluspa Jarpa: Biblioteca de la NO-Historia de Chile." In *Dislocación: Cultural Location and Identity in Times of Globalization.* Ostfildern, Germany: Hatje Cantz Verlag, 2011: 139–47.

Voegelin, Salomé. *Sonic Possible Worlds: Hearing the Continuum of Sound.* New York: Bloomsbury Academic, 2014.

Walcott, Derek. *Collected Poems, 1948–1984.* London: Faber and Faber, 1992.

Walsh, Joan. "The Abu Ghraib Files." *Salon.com,* March 14, 2006. www.salon.com/2006/03/14/introduction_2/ (accessed February 2017).

Warner, Michael. *Publics and Counterpublics.* New York: Zone Books, 2002.

Woodruff, Jeremy. "A Musical Analysis of the People's Microphone. Voices and Echoes in Protest and Sound Art, and *Occupation I* for String Quartet." PhD thesis, University of Pittsburgh, Department of Music, 2014.

Young, James. "The Counter-Monument: Memory against Itself in Germany Today." *Critical Inquiry* 18, Winter 1992: 267–96.

Zurita, Raúl. *Purgatory: Bilingual Edition.* Berkeley: University of California Press, 2009.

Index